HOW TO FAKE A MOON LANDING

Exposing the Myths of Science Denial

DARRYL CUNNINGHAM

INTRODUCTION BY ANDREW C. REVKIN

Abrams ComicArts • New York

“Everyone is entitled to their own opinion; however, everyone is not entitled to their own facts.”

—Michael Specter, staff writer, the *New Yorker* and author of *Denialism: How Irrational Thinking Hinders Scientific Progress, Harms the Planet, and Threatens Our Lives*

(with a nod to Bernard Baruch and Senator Daniel Patrick Moynihan)

Editor: Sheila Keenan
Project Manager: Charles Kochman
Designer: Darryl Cunningham, Sara Corbett
Production Manager: Alison Gervais

Library of Congress Cataloging-in-Publication Data:
Cunningham, Darryl.
How to fake a moon landing : exposing the myths of science denial / Darryl Cunningham.
p. cm.
Includes bibliographical references.
ISBN 978-1-4197-0689-9
1. Pseudoscience—Comic books, strips, etc. I. Title.
Q172.5.P77C86 2013
001.9—dc23
2012042210

Originally published in the UK in 2012 by Myriad Editions under the title *Science Tales: Lies, Hoaxes, and Scams*. “The MMR Vaccination Scandal” was originally titled “The Facts in the Case of Dr. Andrew Wakefield.”

Printed and bound in China
10 9 8 7 6 5 4 3 2 1

Abrams ComicArts books are available at special discounts when purchased in quantity for premiums and promotions as well as fundraising or educational use. Special editions can also be created to specification. For details, contact specialsales@abramsbooks.com or the address below.

ABRAMS
THE ART OF BOOKS SINCE 1949
115 West 18th Street
New York, NY 10011
www.abramsbooks.com

CONTENTS

INTRODUCTION

I was happy to discover Darryl Cunningham's book because he takes a long tradition of demystification and truth seeking to a new generation in a new way. The tradition goes back centuries. Galileo and Darwin fought mightily to use observation, logic, and mathematics to create a reality-based view of the universe and humanity's place in it. It's never been easy. For Galileo, who used observations of the planets and stars to show that the Earth was not the center of the universe, the result was house arrest for the final years of his life as religious leaders fought the erosion of worldviews based on faith. Darwin meticulously and patiently spent more than twenty years assembling his argument and evidence for evolution before the publication of *The Origin of Species*, yet a large portion of society still rejects that firmly established theory.

In recent decades, scientists have been aided by generations of science writers and journalists in their quest to explain and defend their insights about the forces shaping the world. Also helping out are professional debunkers like James Randi. In the early 1970s, at age sixty, Randi moved from a career as a magician (his stage name was the Amazing Randi) to a life focused on separating myth and illusion from reality as a co-founder of the Committee for Scientific Investigation of Claims of the Paranormal. Randi has since been joined by television's popular *MythBusters* team and defenders of the scientific method ranging from Bill Nye the Science Guy to a growing network of science blogs. Just one example is RealClimate.org, a vital touchstone on climate-change research maintained by a team of accomplished climatologists. NASA has gotten into the act using the tools of social media to tamp down occasional waves of hype surrounding a serious challenge—efforts to find and track asteroids that might someday collide with Earth. The space agency's @AsteroidWatch Twitter account is followed by nearly one million people, and provides a beacon of credibility in the great online mash-up of rumors and facts.

Cunningham's innovation is to create a refreshing mix of clear explanatory writing, touches of humor, and engaging and accessible artwork. Like other debunkers, of course, he is more of a truth seeker than a truth teller, given that scientific knowledge is, by nature, a moving target. It was "true" until this century that stomach ulcers were caused by stress or bad diets. After many years of resistance from mainstream medical scientists, it has become clear that a bacterium, *Helicobacter pylori*, is the culprit. (The discoverers had to wait more than twenty years to get their Nobel Prize in 2005.) More recently, biologists studying the human "microbiome"—the human body as an ecosystem—have found that while these bacteria harm the stomach, they provide some benefit to the esophagus, and their decline in Western societies may be related to a rise in asthma, as well. The journey continues.

By fostering the capacity for critical thinking rather than simply providing examples, *How to Fake a Moon Landing* is invaluable given that science and technology will always be providing new material for myth makers to latch on to. The existing debate over nuclear power and genetically modified foods will soon give way to battles over synthetic biology and nanotechnology.

With this book, Cunningham reveals the roots of misinformation and intentional disinformation. There's plenty of both out there. But, more important, he provides a road map for how anyone can do this. One of the great challenges—and opportunities—in twenty-first-century life lies in finding ways to navigate the sea of assertions and facts we face on the Internet and come away with some sense of what's really known and unknown, or simply feared or imagined, in a complicated, dynamic world. When I was coming of age in the mid-twentieth century, there was usually some authority doing the work for you, whether a silver-haired anchorman on the nightly news or the printed front page of a newspaper. Now, no matter what your predispositions, you can probe the web and, seconds later, come away with information or ideas that bolster them.

Maintaining a capacity for critical thinking, including the capacity to appraise one's own beliefs, has never been more important. In some parts of this book, to my eye, Cunningham's deep passion for defrocking falsehoods has carried him too far. I feel he's too hard on chiropractors, for instance, because he doesn't include as rigorous a look at the overselling of many conventional medical treatments for back pain, including surgery, that have not been demonstrated to be more effective than rest and pain medication. I think his portrayal of environmental risks from the gas-drilling method known as "fracking" doesn't adequately account for studies showing that problems, while real, are very rare. But don't take my word for it. Dig in. Find out how to use Google Scholar to probe basic research. Read a scientific paper. Talk to a scientist! They're not hard to find, and most are eager to help the public understand their work.

I'm sure Cunningham would be happy to have readers finish his book and raise their own questions about his conclusions. That would mean they've absorbed his prime message. As he explains in the preface, "This book is pro-science and pro-critical thinking. It isn't a book promoting a scientific élite whom we must all follow, sheep-like."

In the end, I'm sure he's hoping that after you've read this book, you'll be dead set on becoming your own best myth buster. Personally, I can't think of a better outcome.

Andrew Revkin
Garrison, New York
www.nytimes.com/revkin

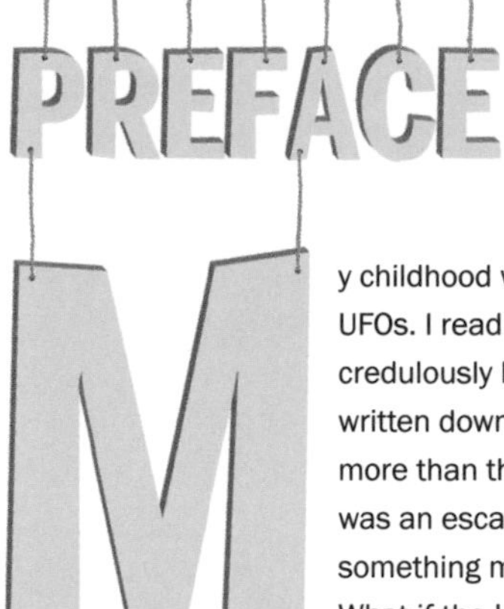

PREFACE

My childhood was filled with fantastical tales of ghosts and UFOs. I read whatever I could on these subjects, rather credulously believing them, simply because they were written down. I powerfully wanted to believe in something more than the ordinary banal world I saw around me. It was an escape from a humdrum home and school life into something more exciting. Why couldn't there be ghosts? What if the Loch Ness monster existed? I searched the skies in the hope that I might see an alien craft, and, disappointingly, never saw one. I craved the fantastic and the bizarre in order to bring color to a monochrome world, only to realize, as I gained a few critical thinking skills, that the evidence for these things just didn't exist.

Of course, the universe has amazing and strange qualities anyway for those who care to see them. There's no need to believe in fantasy in order to see the extraordinary in the world, when reality offers up so much that is astonishing. The journey I took from credulous believer in nonsense to a more hard-headed state of mind was a tough one, in which I had to dispense with much I believed. Accepting that you are wrong in the light of new evidence can be a painful process, but is a necessary one. Science, unlike religion, is in a continuous state of revision, depending on evidence. If you want the truth, then you must go where the facts take you, however uncomfortable that might make you feel. I've argued strongly for the positions I take in the chapters of this book, but I'd like to think I'd be strong enough to change my mind on any of them if the evidence became available. That's a door that must always be kept open.

I've selected these particular subjects because they were the most prominent hot-button science issues at the time of this writing. I had noticed over the past couple of years that whenever I read a science blog or listened to a science podcast, these subjects would come up as the most controversial, time and again. The level of misunderstanding among much of the general public, not just on the issues themselves, but about how the scientific process works, never fails to amaze me. This book is my small attempt to rectify that problem. The order of topics moves from "personal choice" medical issues, such as homeopathy and chiropractic, to larger issues that concern the environment, such as fracking and climate change. The whole structure of the book is meant to build up a case for critical thinking and the scientific process itself.

This book is pro-science and pro-critical thinking. It isn't a book promoting a scientific élite whom we must all follow, sheep-like. It is the scientific process itself I'm promoting here, not the scientific establishment. They are just as capable of being fraudulent, corrupted by politics and money, or just plain wrong as any group of humans engaged in any activity. We know the scientific process can be relied on, because, if it couldn't be, the lightbulb wouldn't work when you switched it on, your mobile phone would be a useless brick, and satellites wouldn't be orbiting the planet.

Science isn't a matter of faith or just another point of view. Good science is testable, reproducible, and stands the test of time. What doesn't work in science falls away, and what remains is the truth.

Darryl Cunningham
Yorkshire, England

ACKNOWLEDGMENTS

THIS BOOK WOULDN'T EXIST without the fine work of the many science writers and journalists whose work I've read in the process of research. So thanks in particular to Phil Plait, Jerry Coyne, Brian Deer, Ben Goldacre, Steven Novella, Simon Singh, and Edzard Ernst.

I'd also like to thank the people who have supported me through the long process of this book's creation: Ian Williams, Scott McCloud, Pádraig Ó Méalóid, Peter Clack, Sue Krekorian, Sarah MacIntyre, Megan Donnelly, John Miers, Jonathan Edwards, Louise Evans, Paul Gravett, Peter Stanbury, Joe Gordon, Dean Haspiel, Simon Fraser, Tom Spurgeon, Nick Abadzis, and Eric Orchard; Corinne Pearlman and all at Myriad; and Sheila Keenan, Sara Corbett, and all at Abrams.

THE
MOON
HOAX

ON JULY 20, 1969, THE APOLLO 11 SPACE FLIGHT LANDED THE FIRST HUMANS ON THE MOON.

THIS, AND SUBSEQUENT MISSIONS, CARRIED OUT BY THE UNITED STATES ARE CONSIDERED A MAJOR ACCOMPLISHMENT IN HUMAN EXPLORATION

AND A U.S. VICTORY IN THE SPACE RACE WITH THE SOVIET UNION.

HOWEVER, IN THE DECADES SINCE, A VOCAL MINORITY HAS VOICED CLAIMS THAT THE MOON LANDINGS

NEVER HAPPENED.

THEY MAKE THE CLAIM THAT THE MOON LANDINGS WERE SHOT ON A FILM SET.
THE EVIDENCE THEY PRESENT DOESN'T AMOUNT TO MUCH AND IS EASILY REFUTED.
THE DENVER POST
Americans Walk on Moon; Lunar Takeoff Successful
'One Small Step for Man . . .'
World Sees Astronauts Plant Flag
'They Came in Peace . . .'
'. . . A Giant Leap for Mankind'
BUT THE MERE FACT THAT THESE ACCUSATIONS KEEP BEING MADE HAS CREATED DOUBT WHERE PREVIOUSLY THERE WAS NONE.
WHY ARE THERE NO STARS IN THE MOON PHOTOS?
THE MOON'S SURFACE PRESENTS A LOT OF GLARE FOR A PHOTOGRAPHER.
THERE'S NO ATMOSPHERE TO DIFFUSE THE SUNLIGHT AND THIS CREATES A LANDSCAPE OF HARSH CONTRASTS.

CAMERA SETTINGS FOR THIS IMAGE WOULD RENDER BACKGROUND STARS TOO FAINT TO BE SEEN.

IF THE SUN IS THE ONLY LIGHT SOURCE, HOW IS IT THAT THE ASTRONAUTS CAN STILL BE SEEN IN THE SHADOW OF THE LANDER?

THERE MUST BE ANOTHER LIGHT SOURCE.

LOOK AT THIS BRIGHT SURFACE.

WE KNOW THIS SURFACE REFLECTS LIGHT, BECAUSE IF IT DIDN'T, WE WOULDN'T BE ABLE TO SEE IT.

THE MOON IS LUMINOUS.

OBJECTS, EVEN IN SHADOW, ARE BATHED IN THIS LIGHT.

IT'S POSSIBLE TO READ A NEWSPAPER BY THE LIGHT OF THE FULL MOON.
The Washington Post
'The Eagle Has Landed'—
Two Men Walk on the Moon

NO OTHER LIGHT SOURCE IS NEEDED.
The Washington Post
'The Eagle Has Landed'— Two Men Walk on the Moon
THE UNITED STATES FLAG RIPPLES AND BENDS, AS IF IN A BREEZE.
HOW IS THIS POSSIBLE WHEN THERE'S NO ATMOSPHERE ON THE MOON?
THE FLAG MOVED ONLY WHEN IT WAS BEING ERECTED. THE MOMENT THE FLAG WAS IN POSITION, IT STOPPED DEAD.
THE FLAG WAS CREATED WITH A RIGID EXTENDABLE SUPPORT PIECE RUNNING ALONG ITS TOP, SO THAT IT WOULD LOOK TAUT.
AT THEIR TECHNICAL DEBRIEFING, THE ASTRONAUTS REPORTED A FEW PROBLEMS WITH THE FLAG DEPLOYMENT.

THEY HAD TROUBLE EXTENDING THE HORIZONTAL TELESCOPING ROD AND COULD NOT PULL IT ALL THE WAY OUT.
THIS GAVE THE FLAG A RIPPLE EFFECT. LATER CREWS INTENTIONALLY LEFT THE ROD PARTIALLY RETRACTED, AS THEY LIKED THE WAY IT LOOKED.
A ROCKET CAPABLE OF LANDING ON THE MOON SHOULD HAVE BURNED A HUGE CRATER ON THE SURFACE.
YET THERE'S NOTHING THERE.
THE LANDER'S POWERFUL ROCKET WAS CAPABLE OF 10,000 POUNDS OF THRUST.
HOWEVER, A REDUCED 3,000 POUNDS WAS ALL THAT WAS NEEDED TO ENSURE A SAFE LANDING.

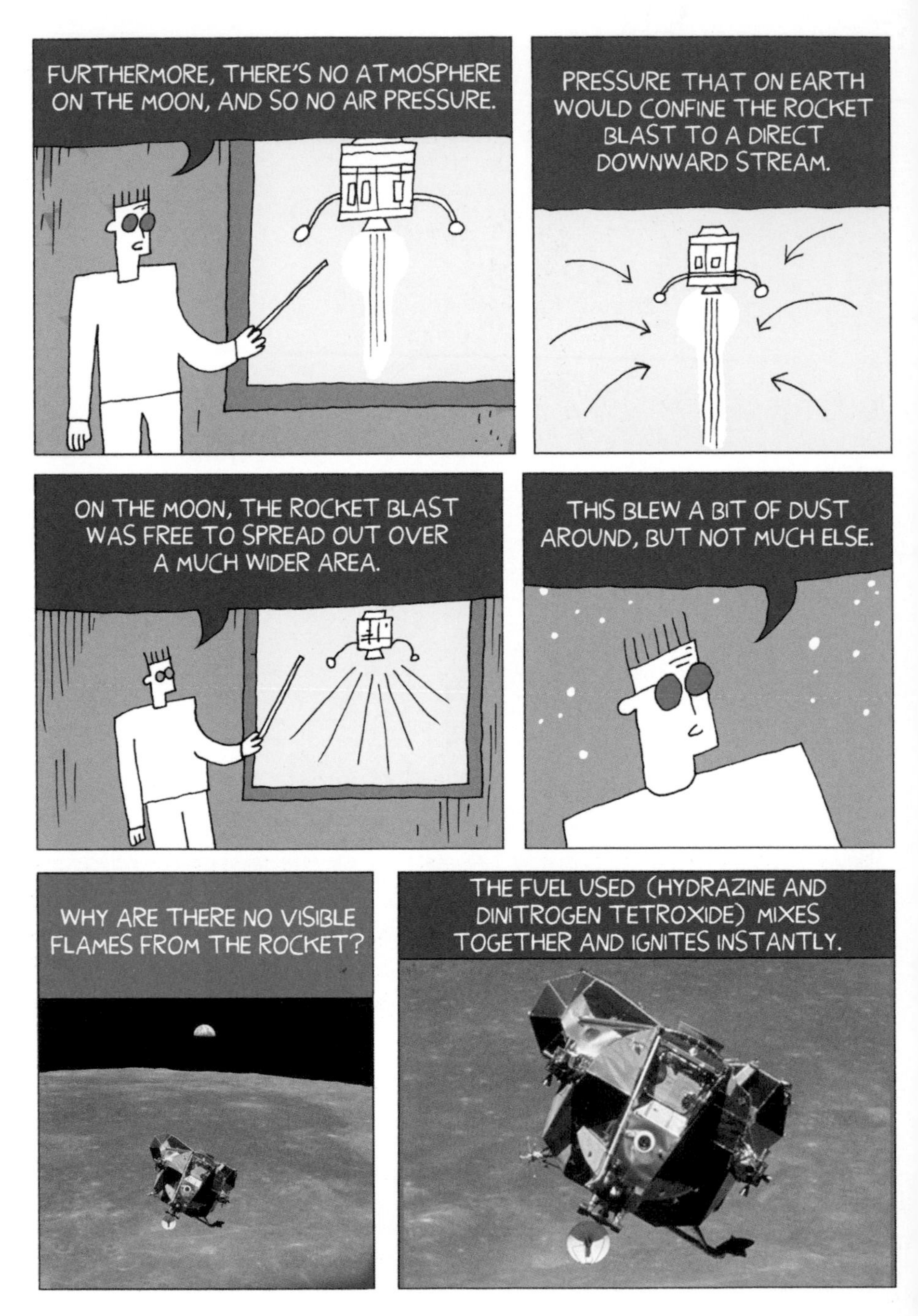
FURTHERMORE, THERE'S NO ATMOSPHERE ON THE MOON, AND SO NO AIR PRESSURE.
PRESSURE THAT ON EARTH WOULD CONFINE THE ROCKET BLAST TO A DIRECT DOWNWARD STREAM.
ON THE MOON, THE ROCKET BLAST WAS FREE TO SPREAD OUT OVER A MUCH WIDER AREA.
THIS BLEW A BIT OF DUST AROUND, BUT NOT MUCH ELSE.
WHY ARE THERE NO VISIBLE FLAMES FROM THE ROCKET?
THE FUEL USED (HYDRAZINE AND DINITROGEN TETROXIDE) MIXES TOGETHER AND IGNITES INSTANTLY.

THIS PRODUCES A FLAME THAT IS COMPLETELY TRANSPARENT.

THIS FUEL IS STILL USED IN SOME ORBITING SPACECRAFT AND IMAGES OF THESE CRAFT SHOW NO FLAMES.

IF YOU RUN THE MOON FOOTAGE AT DOUBLE SPEED, IT LOOKS LIKE IT WAS FILMED ON EARTH.

IN 2008, JAMIE HYNEMAN AND ADAM SAVAGE OF THE POPULAR DISCOVERY CHANNEL SHOW MYTHBUSTERS

TESTED MANY OF THE MOON HOAX CLAIMS. (EPISODE 104. AIR DATE: AUGUST 27TH, 2008).

THEY ATTEMPTED TO DUPLICATE THE ASTRONAUT'S MOON WALK IN EARTH'S GRAVITY.

A HARNESS WAS USED TO MAKE THE WEARER ONE-SIXTH OF HIS NORMAL WEIGHT.
WOO!
THE WEARER'S MOVEMENTS WERE THEN SHOT AT 48 FRAMES A SECOND.
THE TAPE WAS THEN PLAYED BACK AT THE REGULAR 24 PER SECOND.
THE SLOW-MOTION RESULTS, ALTHOUGH QUITE CLOSE TO THE NASA FOOTAGE
STILL LACKED THE SMOOTH LOW-GRAVITY LOOK OF THE ASTRONAUT'S WALK.
THE MYTHBUSTERS THEN WENT ON TO COMPARE THIS RESULT WITH AN ACTUAL DEMONSTRATION IN MICRO-GRAVITY.

MAKING USE OF A MODIFIED BOEING 727-200, THE DUO WERE ABLE TO EXPERIENCE LUNAR GRAVITY ITSELF.
NORTHROP GRU
THIS WAS ACHIEVED BY HAVING THE AIRCRAFT FLY A SERIES OF PARABOLIC ARCS.
ON THE DOWNWARD SIDE OF EACH ARC, THE GRAVITY OF THE CABIN BECAME AN EXACT MATCH
WITH THE MOON'S GRAVITATIONAL PULL.
IN THE SHOW ADAM SAVAGE TALKS ABOUT HIS EXPERIENCE OF WALKING IN MICRO-GRAVITY.
ALL THE NASA FOOTAGE MAKES SENSE TO ME NOW.

COUNT DOWN
First man land on moon

Wapakoneta Daily News

NEIL STEPS ON THE MOON

"All people one," Nixon tells pair

Navy beefs up armada in Pacific

Armstrong, Aldrin set for hazardous return

Mom, dad worries lighter after walk

ADAM SAVAGE'S MOVEMENTS IN MICRO-GRAVITY WERE AN EXACT MATCH WITH THE ASTRONAUT'S WALK.

MAN HAD GONE TO THE MOON.

382 KILOGRAMS OF MOON ROCK AND MOON DUST WERE COLLECTED DURING THE APOLLO MISSIONS. GEOLOGISTS WORLDWIDE HAVE BEEN EXAMINING THESE SAMPLES FOR THIRTY YEARS.

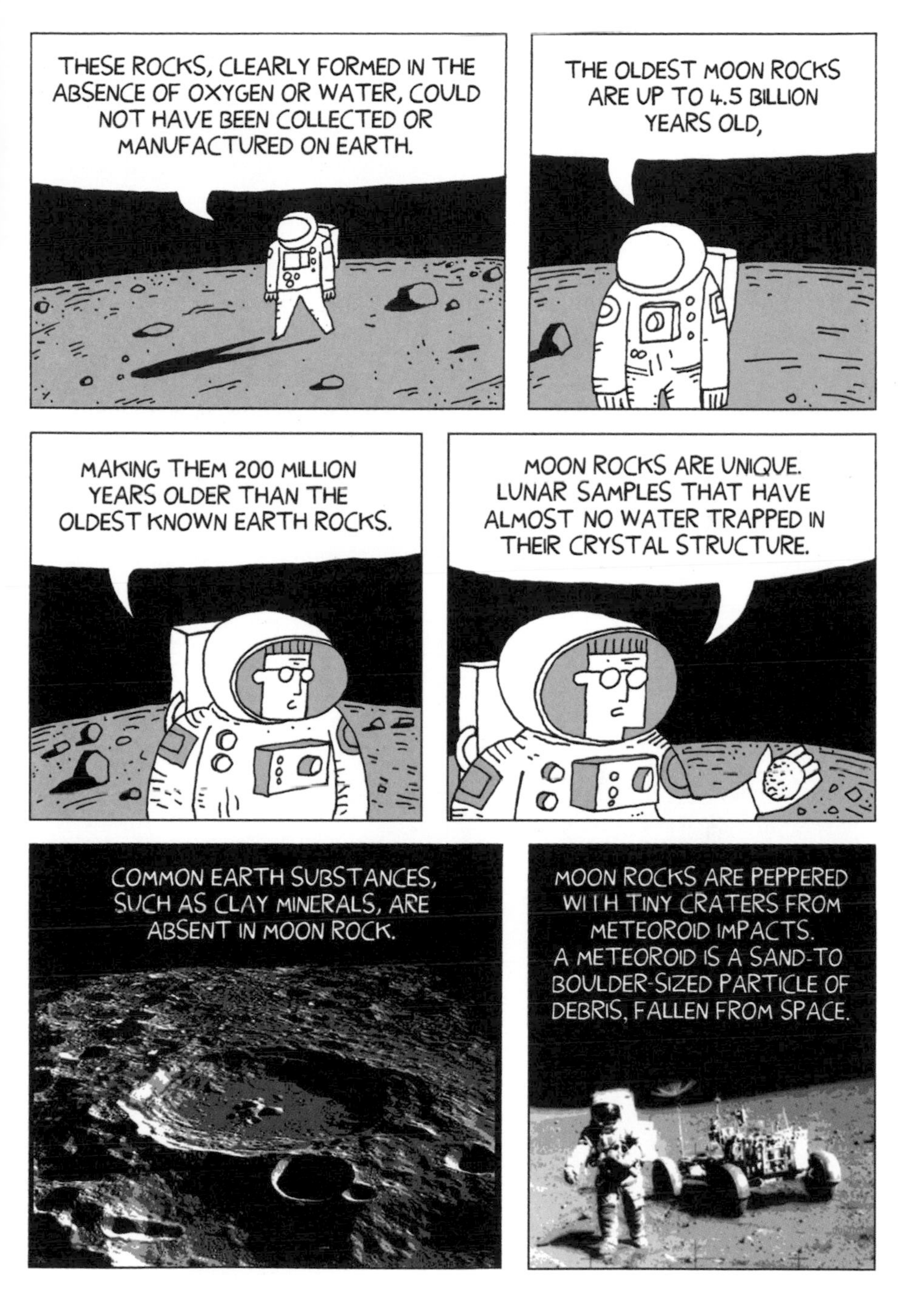
THESE ROCKS, CLEARLY FORMED IN THE ABSENCE OF OXYGEN OR WATER, COULD NOT HAVE BEEN COLLECTED OR MANUFACTURED ON EARTH.
THE OLDEST MOON ROCKS ARE UP TO 4.5 BILLION YEARS OLD,
MAKING THEM 200 MILLION YEARS OLDER THAN THE OLDEST KNOWN EARTH ROCKS.
MOON ROCKS ARE UNIQUE. LUNAR SAMPLES THAT HAVE ALMOST NO WATER TRAPPED IN THEIR CRYSTAL STRUCTURE.
COMMON EARTH SUBSTANCES, SUCH AS CLAY MINERALS, ARE ABSENT IN MOON ROCK.
MOON ROCKS ARE PEPPERED WITH TINY CRATERS FROM METEOROID IMPACTS. A METEOROID IS A SAND-TO BOULDER-SIZED PARTICLE OF DEBRIS, FALLEN FROM SPACE.

THIS COULD ONLY HAPPEN TO ROCKS FROM A PLANET WITH LITTLE OR OR NO ATMOSPHERE
LIKE THE MOON.
I COULD GO ON REBUTTING THE MOON HOAX CLAIMS FOR HUNDREDS OF PAGES.
BUT I KNOW IT WON'T MAKE ANY DIFFERENCE TO THE HARD-CORE BELIEVERS.
THE IDEA THAT THOUSANDS OF ENGINEERS, SCIENTISTS, ASTRONAUTS, AND SUPPORT STAFF HAVE KEPT QUIET AND PERPETRATED THIS HOAX FOR DECADES IS MIND-BOGGLING.
NASA
BUT THERE WILL ALWAYS BE PEOPLE WHO'LL CLAIM A CONSPIRACY FOR JUST ABOUT ANYTHING.
APOLLO 11

REAL CONSPIRACIES ARE MESSY, FLAWED, HUMAN AFFAIRS.
HISTORY HAS SHOWN US REPEATEDLY HOW DIFFICULT IT IS EVEN FOR THE MOST POWERFUL TO KEEP THEIR DUBIOUS ACTIVITIES SECRET.
The New York Times
NIXON RESIGNS
HE URGES A TIME OF 'HEALING'; FORD WILL TAKE OFFICE TODAY
'Sacrifice' Is Praised; Kissinger to Remain
The 37th President Is First to Quit Post
HOW CREDIBLE IS IT, THEN, THAT IN THE WORLD OF THE INTERNET, OF TALK SHOWS, FILM AND BOOK DEALS,
NOT ONE ASTRONAUT OR NASA EMPLOYEE HAS BLOWN THE WHISTLE ON THIS ALLEGED HOAX?
IN ORDER TO FOOL BOTH THE WORLD, AND ITS OWN PEOPLE, THE UNITED STATES GOVERNMENT WOULD HAVE HAD TO ROPE THOUSANDS OF PEOPLE INTO A GLOBAL CONSPIRACY, COSTING BILLIONS.
FAR EASIER TO ACTUALLY GO TO THE MOON. WHICH IS, OF COURSE, WHAT THEY DID.
END

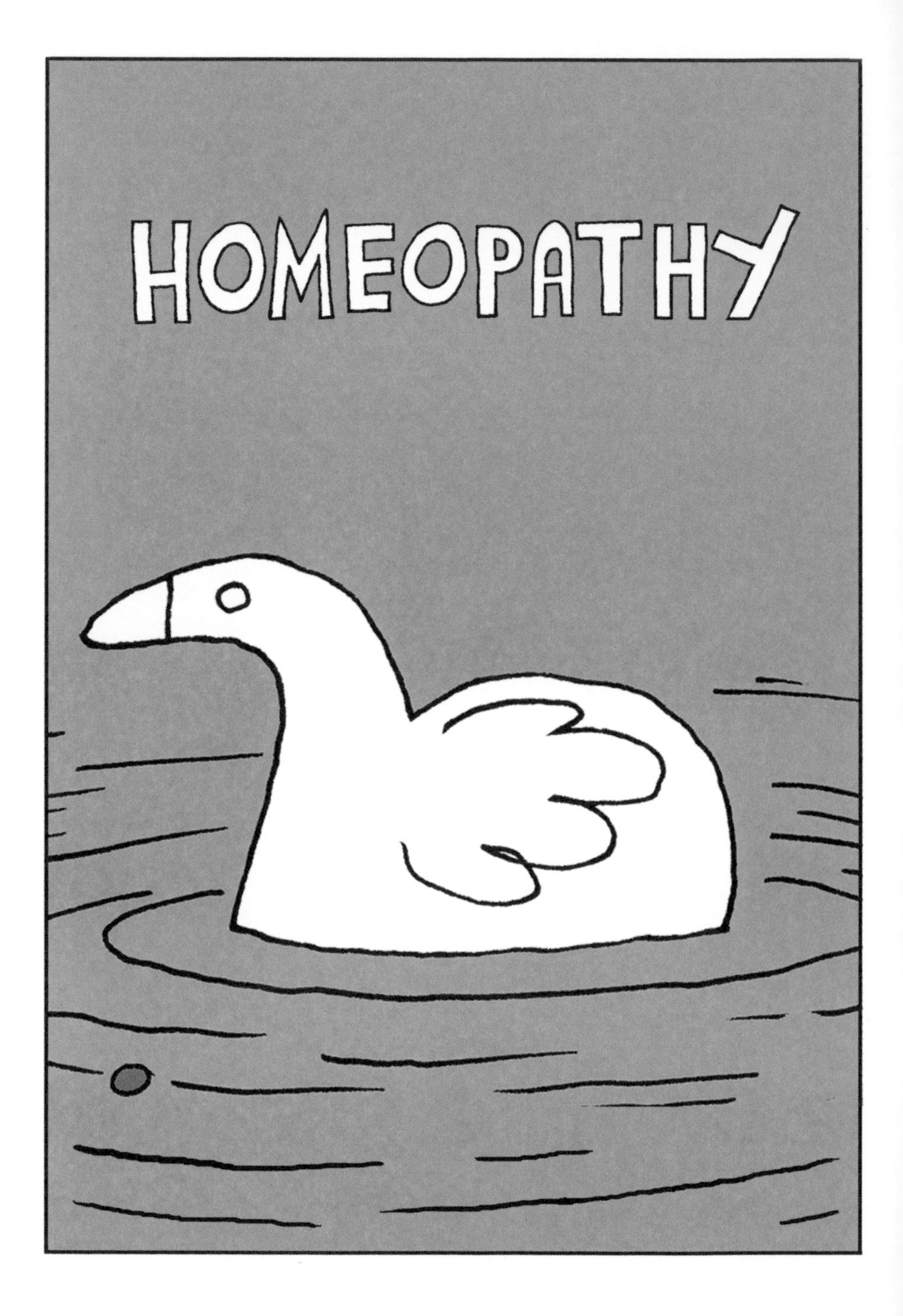
HOMEOPATHY

HOMEOPATHY IS A SYSTEM OF MEDICINE THAT TREATS THE INDIVIDUAL WITH HIGHLY DILUTED SUBSTANCES.
SUBSTANCES THAT ARE GIVEN IN TABLET FORM. THIS IS THOUGHT TO TRIGGER THE BODY'S NATURAL HEALING SYSTEM.
BASED ON A PERSON'S SYMPTOMS, A HOMEOPATH WILL MATCH THE MOST APPROPRIATE MEDICINE TO THE PATIENT.
HOMEOPATHY IS BASED ON TWO MAIN HYPOTHESES. THE FIRST IS THE LAW OF SIMILARS.
THE IDEA THAT ILLNESSES CAN BE CURED BY SMALL DOSES OF SUBSTANCES THAT CAUSE THESE SAME SYMPTOMS.

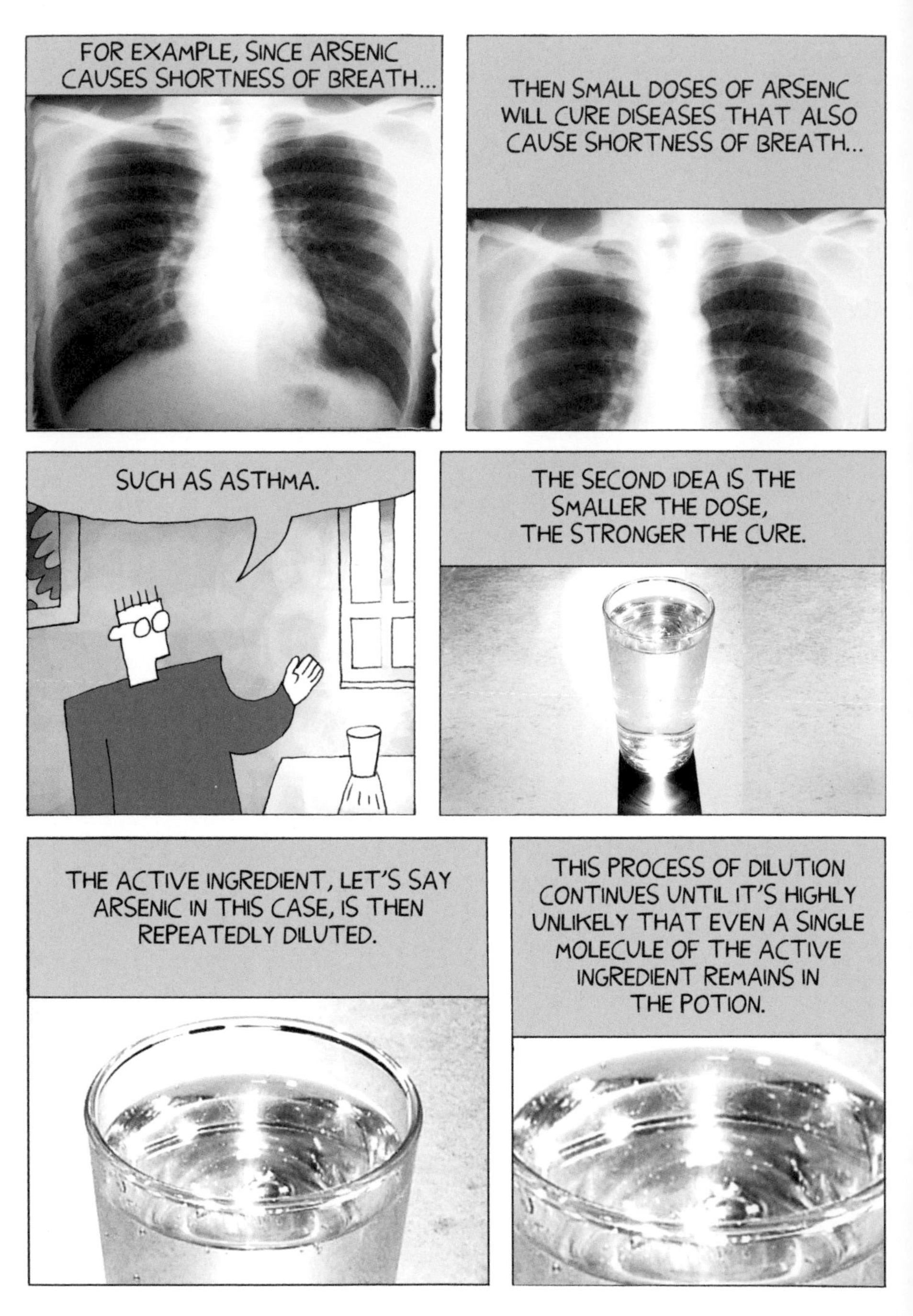
FOR EXAMPLE, SINCE ARSENIC CAUSES SHORTNESS OF BREATH...
THEN SMALL DOSES OF ARSENIC WILL CURE DISEASES THAT ALSO CAUSE SHORTNESS OF BREATH...
SUCH AS ASTHMA.
THE SECOND IDEA IS THE SMALLER THE DOSE, THE STRONGER THE CURE.
THE ACTIVE INGREDIENT, LET'S SAY ARSENIC IN THIS CASE, IS THEN REPEATEDLY DILUTED.
THIS PROCESS OF DILUTION CONTINUES UNTIL IT'S HIGHLY UNLIKELY THAT EVEN A SINGLE MOLECULE OF THE ACTIVE INGREDIENT REMAINS IN THE POTION.

HOWEVER, ACCORDING TO HOMEOPATHIC THEORY, THE DILUTED REMEDY NEEDS NO TRACE OF THE ACTIVE INGREDIENT TO WORK
SINCE THE SUBSTANCE IS IMPRINTED ON THE STRUCTURE OF THE WATER.
EACH TIME THE INGREDIENT IS DILUTED, IT IS VIGOROUSLY SHAKEN.
THIS PROCESS COMPELS THE WATER TO "REMEMBER" THE GRADUALLY VANISHING INGREDIENT.
THE MORE THE REMEDY IS SHAKEN, AND THE LESS OF THE ORIGINAL INGREDIENT THERE IS,
THE MORE POTENT AND POWERFUL THE REMEDY BECOMES.

CAN WATER ACTUALLY HAVE A MEMORY?
IN 1988, JACQUES BENVENISTE, A FRENCH BIOLOGIST, CLAIMED THAT IT COULD.
THE HIGHLY RESPECTED JOURNAL NATURE PUBLISHED HIS CONTROVERSIAL RESEARCH PAPER.
nature
BENVENISTE STARTED WITH A SUBSTANCE THAT CAUSED AN ALLERGIC REACTION.
THIS WAS DILUTED REPEATEDLY UNTIL THERE WAS NOTHING LEFT EXCEPT PURE WATER.
YET THIS WATER STILL MANAGED TO TRIGGER AN ALLERGIC REACTION WHEN ADDED TO LIVING CELLS.

JOHN MADDOX, EDITOR OF NATURE REALIZED THAT BENVENISTE'S RESEARCH WOULD BE CONTROVERSIAL.
SO IT WAS ACCOMPANIED BY A DISCLAIMER.
READERS OF THIS ARTICLE MAY SHARE THE INCREDULITY OF THE MANY REFEREES.
NATURE HAS THEREFORE ARRANGED FOR INDEPENDENT INVESTIGATORS TO OBSERVE THE REPETITIONS OF THE EXPERIMENT.
THE INVESTIGATION TEAM WAS LED BY MADDOX HIMSELF. HE WAS JOINED BY CHEMIST WALTER STEWART
AND JAMES RANDI, A MAGICIAN KNOWN FOR HIS EXPERTISE IN INVESTIGATING EXTRAORDINARY CLAIMS.

WHEN THE BENVENISTE TEAM REPEATED THE EXPERIMENT, THE INVESTIGATORS WENT TO EXTREME LENGTHS TO ENSURE THAT NONE OF THE SCIENTISTS KNEW WHICH WERE THE HOMEOPATHIC SOLUTIONS
AND WHICH WERE THE SAMPLES.
THEY WENT SO FAR AS TO TAPE SAMPLES TO THE CEILING, SO THAT ANY ATTEMPT TO TAMPER WITH IT OVERNIGHT WOULD BE NOTICED.
THE INVESTIGATORS SOON DISCOVERED THAT THE RESULTS IN BENVENISTE'S LABORATORY WERE UNRELIABLE.
THEY FOUND A MIXTURE OF LOOSE OR NON-EXISTENT CONTROLS, POSSIBLE EQUIPMENT CONTAMINATION, DATA MANIPULATION, AND DATA SELECTION (SELECTING FOR POSITIVE RESULTS AND IGNORING THE NEGATIVE ONES).
WE BELIEVE THAT EXPERIMENTAL DATA HAVE BEEN UNCRITICALLY ASSESSED AND THEIR IMPERFECTIONS INADEQUATELY REPORTED.

DESPITE NATURE MAGAZINE'S DAMNING REPORT, BENVENISTE CONTINUED TO MAINTAIN THAT HIS RESEARCH WAS VALID.
HE LATER FOUNDED A COMPANY, CALLED DIGIBIO, WHICH CLAIMED THAT NOT ONLY DID WATER HAVE A MEMORY,
BUT THAT THIS MEMORY COULD BE DIGITIZED, TRANSMITTED VIA EMAIL, AND THEN REINTRODUCED INTO WATER.
THERE HAVE BEEN MANY ATTEMPTS TO REPRODUCE BENVENISTE'S EXPERIMENTS. BUT ANY POSITIVE RESULTS HAVE BEEN NEITHER CONSISTENT NOR CONVINCING.
ALL EVIDENCE POINTS TO HOMEOPATHIC REMEDIES BEING INERT AND NO MORE EFFECTIVE THAN A PLACEBO
OR JUST LETTING THE ILLNESS RUN ITS COURSE.

ARE THERE ANY ASPECTS OF HOMEOPATHY THAT CAN BE SAID TO WORK?
READER'S VOICE
YES. A HOMEOPATH WILL SPEND AT LEAST AN HOUR, SOMETIMES LONGER,
ASKING DETAILED QUESTIONS ABOUT THE PATIENT'S CURRENT HEALTH, MEDICAL HISTORY, AND LIFESTYLE.
IT'S KNOWN THAT STRESS AND ANXIETY CAN ENHANCE ILLNESS.
SO ANYTHING THAT DIMINISHES STRESS AND ANXIETY, SUCH AS THE ATTENTION A HOMEOPATH GIVES,
?
MAY WELL TRANSLATE INTO BENEFICIAL PHYSIOLOGICAL EFFECTS FOR THE PATIENT.

PATIENTS OF HOMEOPATHY FEEL THAT THEY ARE BEING TREATED AS INDIVIDUALS.
A HOMEOPATHIC CONSULTATION GIVES THE PATIENT A CHANCE TO TALK AT LENGTH ABOUT HIS OR HER PROBLEMS
BLAH, BLAH, BLAH, BLAH, BLAH, BLAH, BLAH, BLAH, BLAH, BLAH, BLAH, BLAH, BLAH, BLAH, BLAH, BLAH, BLAH, BLAH. BLAH, BLAH!
TO A SYMPATHETIC LISTENER, IN A STRUCTURED ENVIRONMENT.
DO GO ON.
THIS IN ITSELF IS THERAPEUTIC. SO, TO ANSWER YOUR QUESTION, THE ASPECT OF HOMEOPATHY THAT DOES WORK
IS ACTUALLY A FORM OF PSYCHOTHERAPY.
I SEE.
READER'S VOICE

WHAT THEN IS THE HARM IN HOMEOPATHY IF THE REMEDIES ARE INERT
AND THE CONSULTATIONS HAVE A POSITIVE THERAPEUTIC EFFECT?
SURELY THERE CAN'T BE ANY SIDE EFFECTS ATTACHED TO THE PRACTICE?
THE MAIN DANGER OF HOMEOPATHY IS THAT IT CAN DISCOURAGE PEOPLE FROM GETTING REAL TREATMENT.
HOMEOPATHY IS NEVER GOING TO CURE A YEAST INFECTION, ASTHMA, OR CANCER.
AN AVOIDANCE OF SCIENCE-BASED MEDICINE CAN ONLY LEAD TO SICKNESS

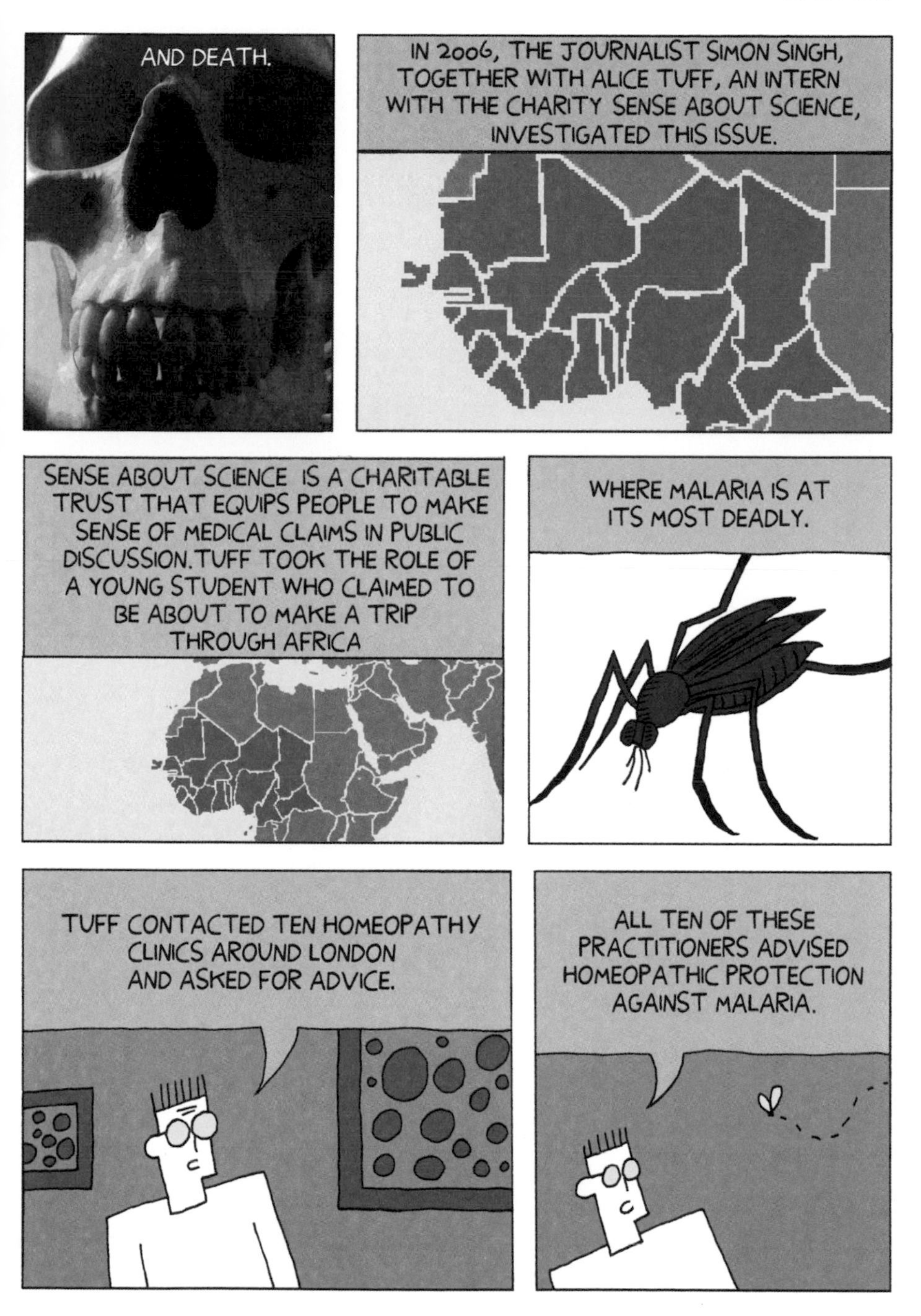
AND DEATH.
IN 2006, THE JOURNALIST SIMON SINGH, TOGETHER WITH ALICE TUFF, AN INTERN WITH THE CHARITY SENSE ABOUT SCIENCE, INVESTIGATED THIS ISSUE.
SENSE ABOUT SCIENCE IS A CHARITABLE TRUST THAT EQUIPS PEOPLE TO MAKE SENSE OF MEDICAL CLAIMS IN PUBLIC DISCUSSION. TUFF TOOK THE ROLE OF A YOUNG STUDENT WHO CLAIMED TO BE ABOUT TO MAKE A TRIP THROUGH AFRICA
WHERE MALARIA IS AT ITS MOST DEADLY.
TUFF CONTACTED TEN HOMEOPATHY CLINICS AROUND LONDON AND ASKED FOR ADVICE.
ALL TEN OF THESE PRACTITIONERS ADVISED HOMEOPATHIC PROTECTION AGAINST MALARIA.

NONE OF THEM RECOMMENDED TAKING BOTH THE CONVENTIONAL TREATMENT AND THE HOMEOPATHIC REMEDIES.
VERY DANGEROUS FOR TUFF, IF SHE HAD ACTUALLY INTENDED TO GO TO WEST AFRICA.
IN 2005, THE HEALTH PROTECTION AGENCY, AN INDEPENDENT UK ORGANIZATION, ISSUED A WARNING
Health Protection Agency
DUE TO THE NUMBER OF PEOPLE FALLING ILL AFTER USING HOMEOPATHIC REMEDIES.
THERE IS NO SCIENTIFIC PROOF THAT HOMEOPATHIC REMEDIES ARE EFFECTIVE IN PREVENTING OR TREATING MALARIA.
HPA SPOKESPERSON

THERE ARE ALSO DANGERS WHEN HOMEOPATHS REPLACE DOCTORS AS SOURCES OF MEDICAL ADVICE.
IN 2002, EDZARD ERNST AND KATJA SCHMIDT OF THE UNIVERSITY OF EXETER
CONDUCTED A REVEALING SURVEY AMONG UK HOMEOPATHS.
THEY EMAILED 168 HOMEOPATHS AND PRETENDED TO BE A MOTHER
QUACK! QUACK!
ASKING FOR ADVICE ON WHETHER SHE SHOULD VACCINATE HER CHILD AGAINST MEASLES, MUMPS, AND RUBELLA.
OF THE 77 HOMEOPATHS THAT REPLIED, ONLY TWO ADVISED THE MOTHER TO IMMUNIZE.

PENELOPE DINGLE OF PERTH, AUSTRALIA, WAS DIAGNOSED WITH COLON CANCER IN 2003. HER DOCTORS GAVE HER A GOOD CHANCE OF SURVIVAL WITH STANDARD TREATMENT:

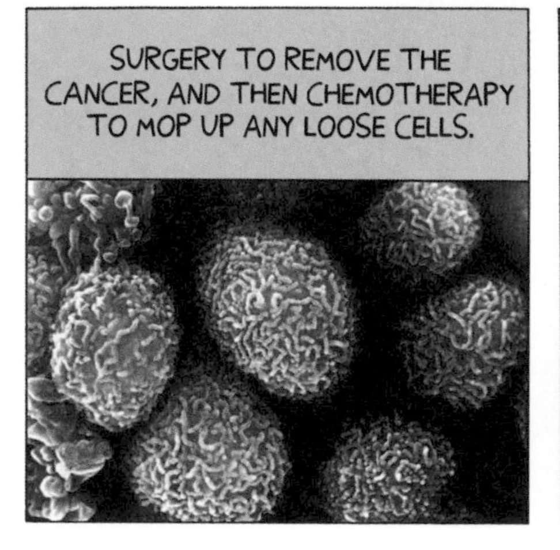

HOWEVER, PENELOPE DINGLE CHOSE TO REFUSE ALL SCIENCE-BASED MEDICAL CARE.

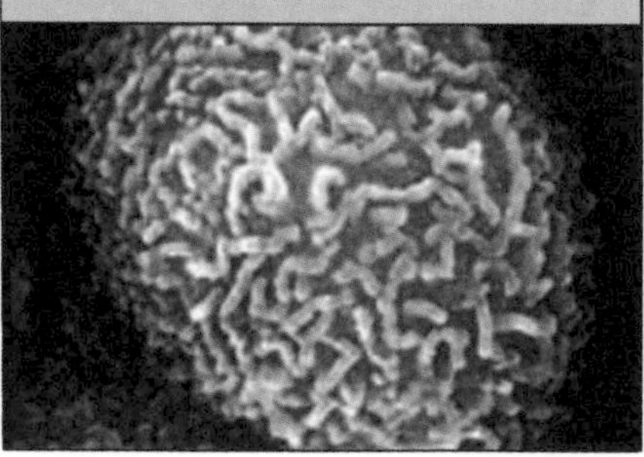

SHE TRUSTED IN HOMEOPATHY AND NUTRITIONAL SUPPLEMENTS INSTEAD.
HER DIARIES FROM THIS TIME SHOW THAT SHE SAW THE CANCER AS A TEST OF HER FAITH IN ALTERNATIVE MEDICINE.
IN OCTOBER 2003, DEBORAH COMBES, A REGISTERED NURSE AND FAMILY FRIEND
PEN DINGLE
WAS ASKED BY PEN DINGLE'S SISTERS TO GO AND HELP.
SHE FOUND PEN DINGLE SWEATY, BREATHLESS, AND EMACIATED.
HER EYES WERE SUNK INTO HER SKULL AND SHE WAS WRITHING IN PAIN, SCREAMING AND VERY FRIGHTENED.

SHE WAS RUSHED INTO THE HOSPITAL AND HAD THE TUMOR REMOVED. HOWEVER, THE CANCER HAD ALREADY SPREAD.
DINGLE DIED TWO YEARS LATER. SHE WAS 45.
PENELOPE DINGLE'S HUSBAND,
DR. PETER DINGLE, AN ASSOCIATE PROFESSOR IN HEALTH AND ENVIRONMENT,
IS SOMETHING OF A MEDIA CELEBRITY IN AUSTRALIA, AN EXPERT ON DIET AND NUTRITION, LIFESTYLE, AND ENVIRONMENTAL HEALTH.
HE HAS A BACHELOR OF EDUCATION IN SCIENCE, A BACHELOR OF ENVIRONMENTAL SCIENCE WITH FIRST CLASS HONORS, AND A PHD.

DR. DINGLE HAS PERHAPS BEEN UNFAIRLY PORTRAYED IN THE MEDIA AS A MAN WHO WAS WILLING TO PLACE HIS WIFE AT RISK IN ORDER TO PUT HIS NUTRITIONAL THEORIES TO THE TEST
ACTUAL BOOK
MY DOG EATS BETTER THAN YOUR KIDS. DR. DINGLE
AND FURTHER HIS OWN CAREER.
MY DOG EATS BETTER THAN YOUR KIDS. DR. DINGLE
HE HAS REPEATEDLY DENIED HAVING ANY REAL INFLUENCE OVER HIS WIFE.
HOWEVER, THERE HAVE BEEN ACCUSATIONS FROM PENELOPE DINGLE'S SISTERS
THAT DR. DINGLE WAS PLANNING A BOOK IN COLLABORATION WITH PEN'S HOMEOPATH, FRANCINE SCRAYEN,
WHICH WOULD HAVE BEEN ABOUT THE CURING OF PEN'S CANCER USING ONLY ALTERNATIVE MEDICINE AND NUTRITION.

THE CORONER'S REPORT WAS DAMNING OF BOTH DR. DINGLE AND FRANCINE SCRAYEN. PENELOPE DINGLE HAD BEEN GIVEN CLEAR AND RELIABLE INFORMATION BY HER OWN DOCTOR,
BUT SHE HAD BEEN INFLUENCED BY MISINFORMATION AND BAD SCIENCE FROM OTHERS.
THE CORONER OBSERVED THAT MRS. SCRAYEN WAS NOT A COMPETENT HEALTH PROFESSIONAL AND THAT SHE HAD ONLY A MINIMAL UNDERSTANDING OF RELEVANT HEALTH ISSUES. UNFORTUNATELY THIS DID NOT STOP HER FROM TREATING THE DECEASED AS A PATIENT.
THE CORONER ALSO NOTED THAT THE RELATIONSHIP BETWEEN THE DECEASED AND MRS. SCRAYEN WENT FAR BEYOND THE NORMAL PATIENT AND HEALTH PROVIDER RELATIONSHIP.
PENELOPE DINGLE HAD BECOME EXTREMELY DEPENDENT ON MRS. SCRAYEN.

IN ORDER FOR HOMEOPATHY TO BE EFFECTIVE IT WOULD HAVE TO WORK IN VIOLATION OF THE PRINCIPLES OF BIOLOGY, CHEMISTRY, AND PHYSICS.
HOMEOPATHY HAS NOTHING TO DO WITH SCIENCE AND EVERYTHING TO DO WITH FAITH. ITS DEVOTEES DON'T CONCERN THEMSELVES WITH EVIDENCE.
THAT SCIENCE DOESN'T RECOGNIZE HOMEOPATHY IS, TO ITS SUPPORTERS, VERY MUCH IN ITS FAVOR.
PENELOPE DINGLE BELIEVED IN THE POWER OF HOMEOPATHY.
SHE PUT HER LIFE AT RISK IN ORDER TO PROVE THAT SCIENCE-BASED MEDICINE WAS UNNECESSARY.
AND THEN SHE DIED.
SIGH!
END

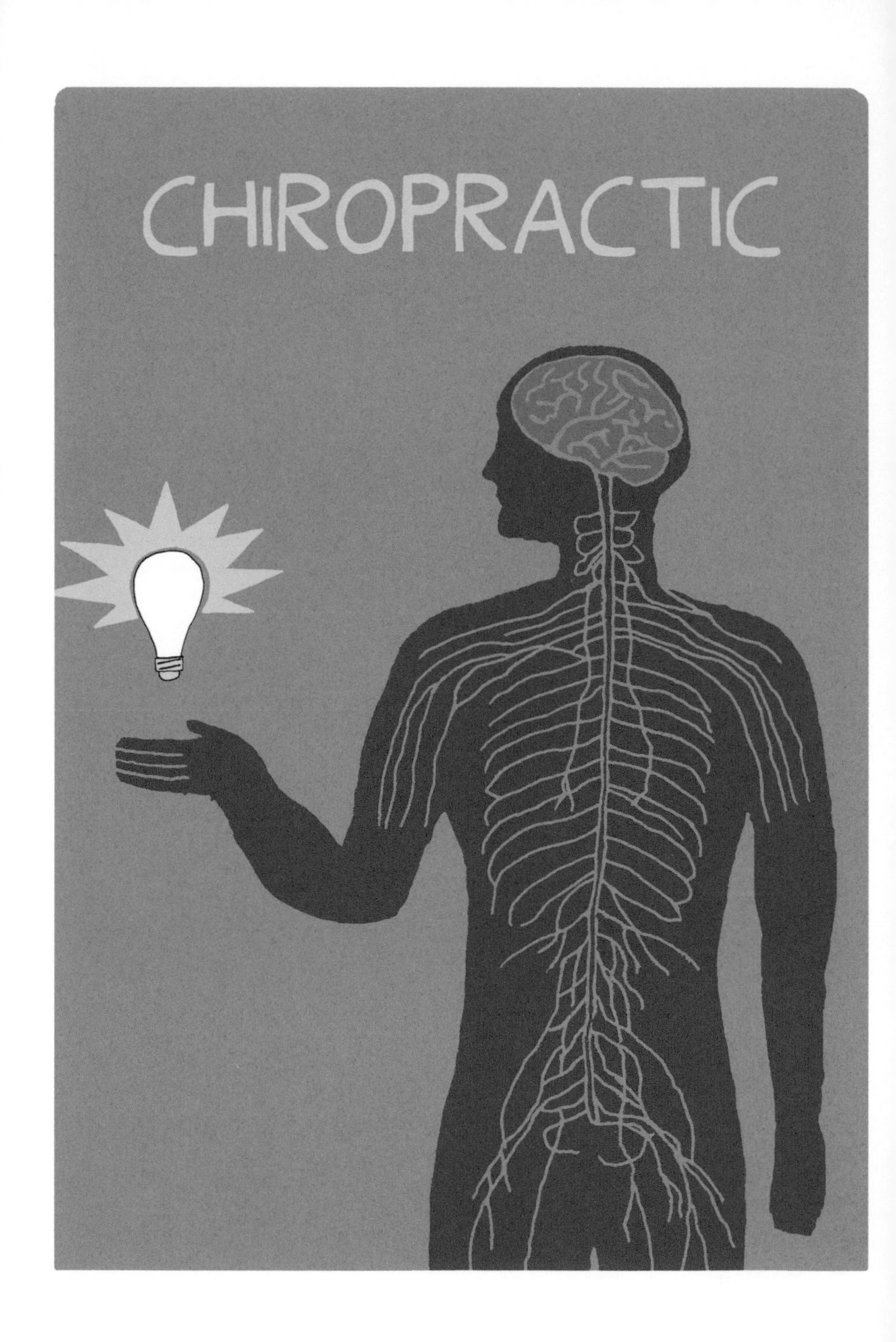
CHIROPRACTIC

OW!
MOST PEOPLE WILL KNOW THAT THE HALLMARK TREATMENT OF A CHIROPRACTOR IS SPINAL MANIPULATION.
ARRGH!
VERY GOOD FOR THE BACK, YOU MIGHT THINK?
MUCH LIKE HOMEOPATHY, CHIROPRACTIC TREATMENT EMERGED OUT OF THE NINETEENTH CENTURY...
A TIME THAT WAS FULL OF ODDBALL IDEAS ABOUT HEALTH CARE.
THE TRADITION WAS FOUNDED BY DANIEL DAVID PALMER. BORN IN TORONTO, CANADA, IN 1845.

PALMER, WHO HAD NO FORMAL MEDICAL TRAINING,

WAS A DEVOTEE OF PHRENOLOGY, MAGNETIC THERAPY, AND OTHER PSEUDOSCIENTIFIC IDEAS OF THE ERA.

HIS INTEREST IN SPINAL MANIPULATION BEGAN IN 1895

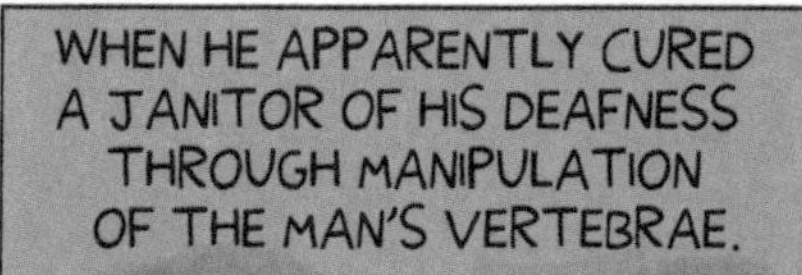
WHEN HE APPARENTLY CURED A JANITOR OF HIS DEAFNESS THROUGH MANIPULATION OF THE MAN'S VERTEBRAE.

HELLO!

PALMER THEN BECAME CONVINCED THAT SPINAL MANIPULATION COULD TREAT ALL MANNER OF HUMAN ILLS.
NO NEED TO SHOUT.

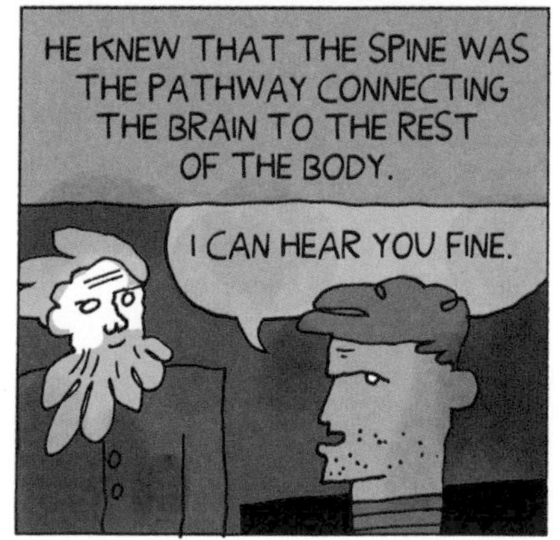
HE KNEW THAT THE SPINE WAS THE PATHWAY CONNECTING THE BRAIN TO THE REST OF THE BODY.
I CAN HEAR YOU FINE.

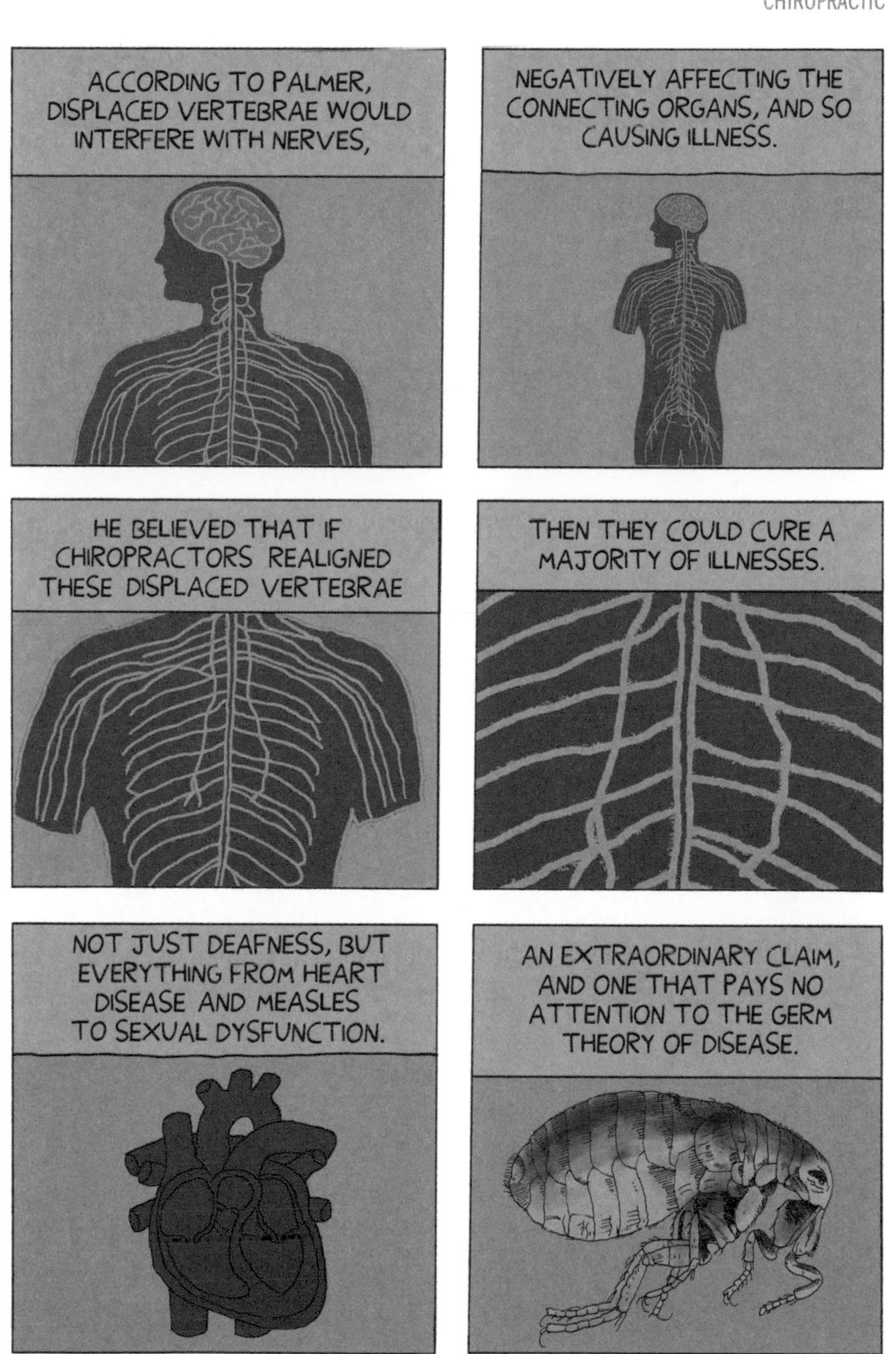
ACCORDING TO PALMER, DISPLACED VERTEBRAE WOULD INTERFERE WITH NERVES,
NEGATIVELY AFFECTING THE CONNECTING ORGANS, AND SO CAUSING ILLNESS.
HE BELIEVED THAT IF CHIROPRACTORS REALIGNED THESE DISPLACED VERTEBRAE
THEN THEY COULD CURE A MAJORITY OF ILLNESSES.
NOT JUST DEAFNESS, BUT EVERYTHING FROM HEART DISEASE AND MEASLES TO SEXUAL DYSFUNCTION.
AN EXTRAORDINARY CLAIM, AND ONE THAT PAYS NO ATTENTION TO THE GERM THEORY OF DISEASE.

PALMER CALLED THESE SPINAL DISPLACEMENTS SUBLUXATIONS.

IN HIS THEORY, SUBLUXATIONS BLOCKED THE BODY'S INNATE INTELLIGENCE,

THE GUIDING ENERGY, CARRYING BOTH PHYSICAL AND PSYCHOLOGICAL SIGNIFICANCE.

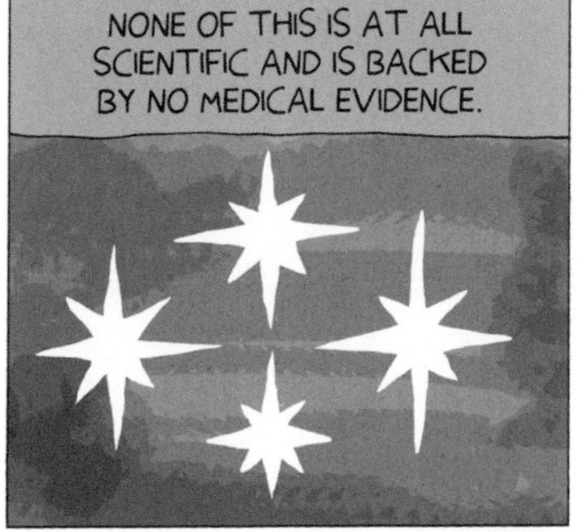
NONE OF THIS IS AT ALL SCIENTIFIC AND IS BACKED BY NO MEDICAL EVIDENCE.

THERE WAS ALWAYS SOMETHING OF THE GURU ABOUT DAVID PALMER,

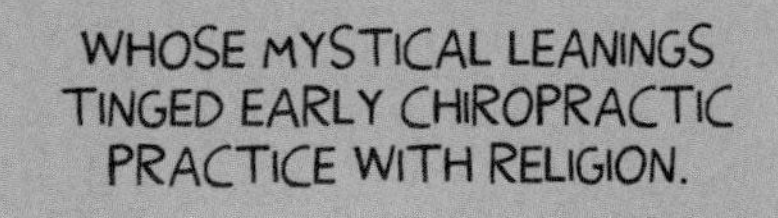
WHOSE MYSTICAL LEANINGS TINGED EARLY CHIROPRACTIC PRACTICE WITH RELIGION.

I AM THE FOUNDER OF CHIROPRACTIC, IN ITS SCIENCE, IN ITS ART, IN ITS PHILOSOPHY, AND IN ITS RELIGION.
CHIROPRACTIC THERAPY HAD ITS FIRST MARTYR, WHEN IN 1906, PALMER WAS JAILED FOR PRACTICING MEDICINE WITHOUT A LICENSE,
AN EVENT THAT ONLY STRENGTHENED THE FAST-GROWING MOVEMENT.
BAH!
DAVID PALMER'S SON WAS BARTLETT JOSHUA PALMER.
B.J. PALMER TOOK OVER THE CHIROPRACTIC MOVEMENT AFTER HIS FATHER'S DEATH.
I'M IN CHARGE NOW.
THE ELDER PALMER DIED THREE WEEKS AFTER A STRANGE INCIDENT, WHEN

HIS SON ACCIDENTALLY RAN OVER HIM WITH A CAR.
ARRGH!

THE OFFICIAL CAUSE OF DEATH WAS TYPHOID, BUT BEING RUN OVER COULDN'T HAVE HELPED.

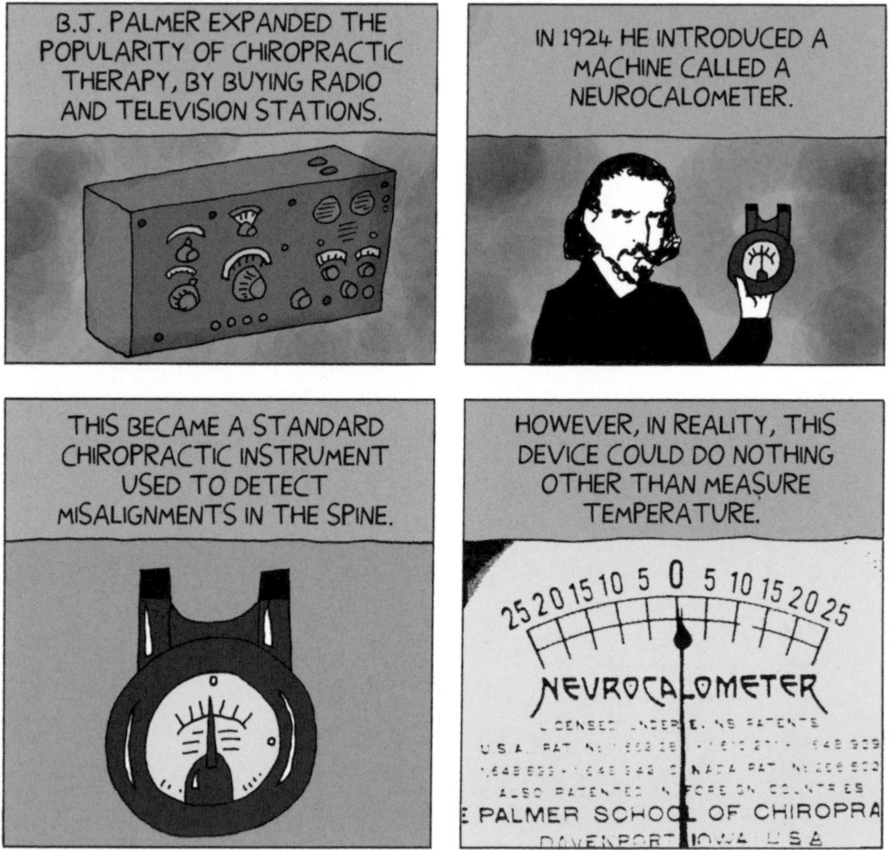
B.J. PALMER EXPANDED THE POPULARITY OF CHIROPRACTIC THERAPY, BY BUYING RADIO AND TELEVISION STATIONS.
IN 1924 HE INTRODUCED A MACHINE CALLED A NEUROCALOMETER.
THIS BECAME A STANDARD CHIROPRACTIC INSTRUMENT USED TO DETECT MISALIGNMENTS IN THE SPINE.
HOWEVER, IN REALITY, THIS DEVICE COULD DO NOTHING OTHER THAN MEASURE TEMPERATURE.
25 20 15 10 5 0 5 10 15 20 25
NEVROCALOMETER
PALMER SCHOOL OF CHIROPRA

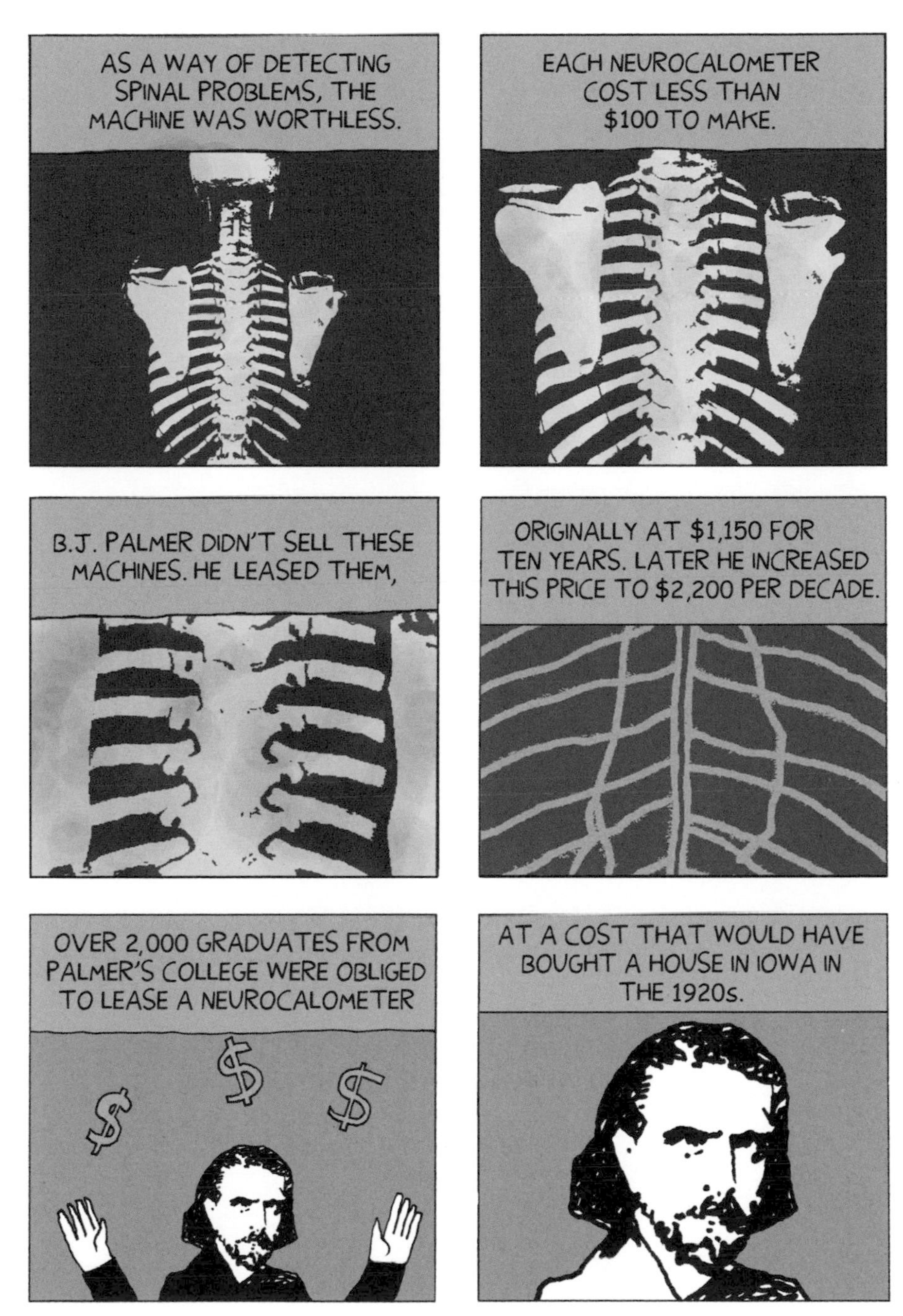
AS A WAY OF DETECTING SPINAL PROBLEMS, THE MACHINE WAS WORTHLESS.
EACH NEUROCALOMETER COST LESS THAN $100 TO MAKE.
B.J. PALMER DIDN'T SELL THESE MACHINES. HE LEASED THEM,
ORIGINALLY AT $1,150 FOR TEN YEARS. LATER HE INCREASED THIS PRICE TO $2,200 PER DECADE.
OVER 2,000 GRADUATES FROM PALMER'S COLLEGE WERE OBLIGED TO LEASE A NEUROCALOMETER
$
$
$
AT A COST THAT WOULD HAVE BOUGHT A HOUSE IN IOWA IN THE 1920s.

IN MODERN TIMES, CHIROPRACTIC HAS BECOME FRAUGHT WITH INTERNAL SCHISMS.

THERE IS A HUGE RANGE OF DIFFERENCE BETWEEN INDIVIDUAL CHIROPRACTORS.

THE MOVEMENT CAN BE DIVIDED INTO THREE GROUPS

STRAIGHTS, MIXERS, AND REFORMERS.

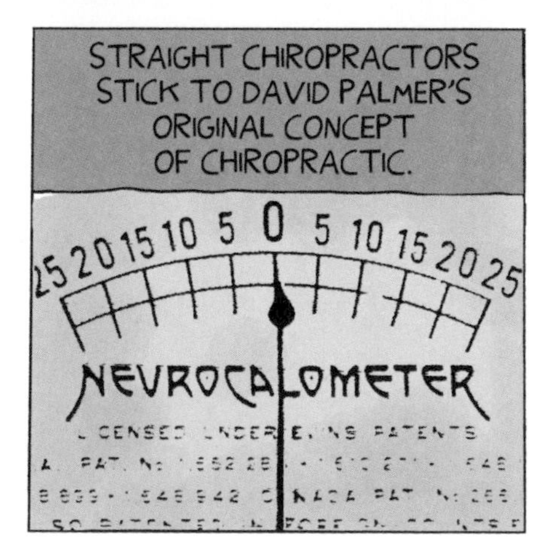
STRAIGHT CHIROPRACTORS STICK TO DAVID PALMER'S ORIGINAL CONCEPT OF CHIROPRACTIC.
25 20 15 10 5 0 5 10 15 20 25
NEVROCALOMETER

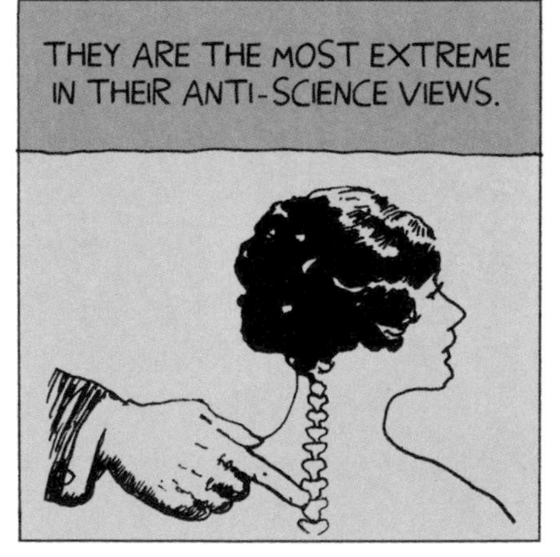
THEY ARE THE MOST EXTREME IN THEIR ANTI-SCIENCE VIEWS.

THEY ADVOCATE A PHILOSOPHICAL RATHER THAN A SCIENTIFIC BASIS FOR HEALTH CARE.
THEY REFER TO PHYSICIANS AS DRUG-PUSHERS AND DISPARAGE THE USE OF SURGERY.
TAKE THESE.
WOW!
THEY ARE CAREFUL NOT TO GIVE DISEASES NAMES, BUT CLAIM TO CURE THEM ANYWAY.
THEY OPPOSE VACCINATIONS.
MIXERS ARE THE LARGEST OF THE THREE GROUPS. MIXERS ACCEPT THAT SOME DISEASES ARE CAUSED BY INFECTION.
THEY HAVE ADAPTED A MORE DEFINED JOB DESCRIPTION, AS BACK SPECIALISTS.
YOW!

MIXERS MAY AT FIRST GLANCE APPEAR MORE RATIONAL, BUT THE HEALTH CARE THEY PRACTICE ALONGSIDE CHIROPRACTIC
TENDS TO BE UNPROVEN AND UNSCIENTIFIC. THEY USE METHODS SUCH AS...
ACUPUNCTURE, THERAPEUTIC TOUCH, HOMEOPATHY, AND HERBAL REMEDIES.
MIXERS DIAGNOSE USING IRIDOLOGY, CONTOUR ANALYSIS, AND APPLIED KINESIOLOGY.
THYROID
L
THIS APPROACH SHOWS AN ANTI-SCIENCE ATTITUDE AND A LACK OF SCIENTIFIC KNOWLEDGE.
THIS GROUP WOULD LIKE TO BE ACCEPTED BY MAINSTREAM MEDICINE.

MIXERS DON'T OPENLY OPPOSE IMMUNIZATION, AS STRAIGHT CHIROPRACTORS DO.
THEY BELIEVE THAT PEOPLE SHOULD HAVE A CHOICE WHETHER TO IMMUNIZE OR NOT.
INFECTION OR PROTECTION? WHICH SHOULD IT BE? HMM!
THIS IGNORES THE FACT THAT IMMUNIZATION IS INEFFECTIVE UNLESS THE MAJORITY OF THE POPULATION IS IMMUNIZED.
MIXERS ASPIRE TO BE A GATEWAY PROFESSION FOR PATIENTS INTO HEALTH CARE.
EVEN THOUGH THEY LACK TRAINING IN GENERAL MEDICAL DIAGNOSIS.
A THIRD AND TINY FRACTION OF CHIROPRACTORS EXIST, WHO SEE THEMSELVES AS REFORMERS.

AND HAVE CALLED FOR A REJECTION OF THE SUBLUXATION THEORY OF ILLNESS.

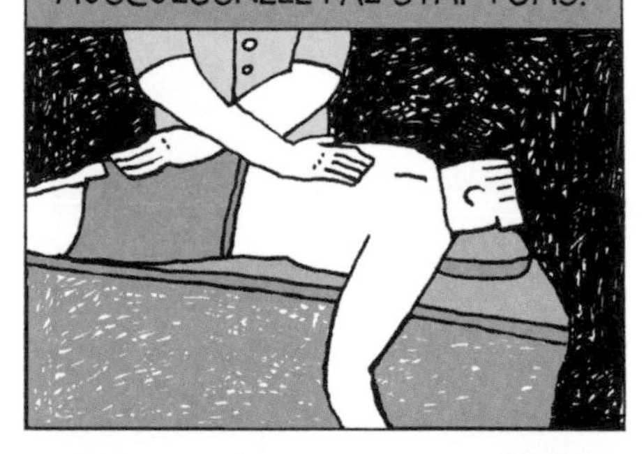

APART FROM ITS ANTI-SCIENCE STANCE, CHIROPRACTIC HAS BEEN CRITICIZED FOR OTHER DUBIOUS ACTIVITIES.

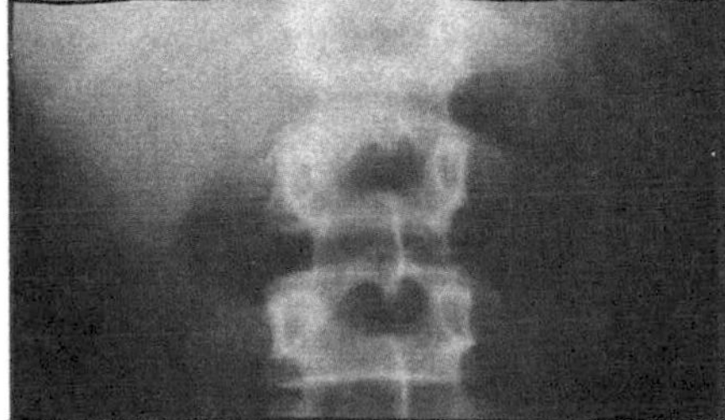

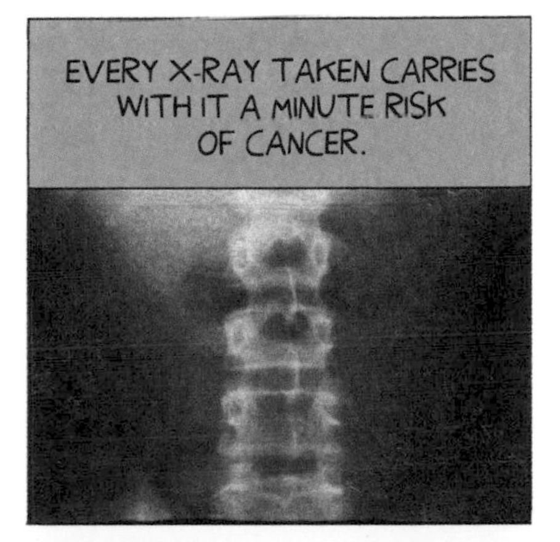

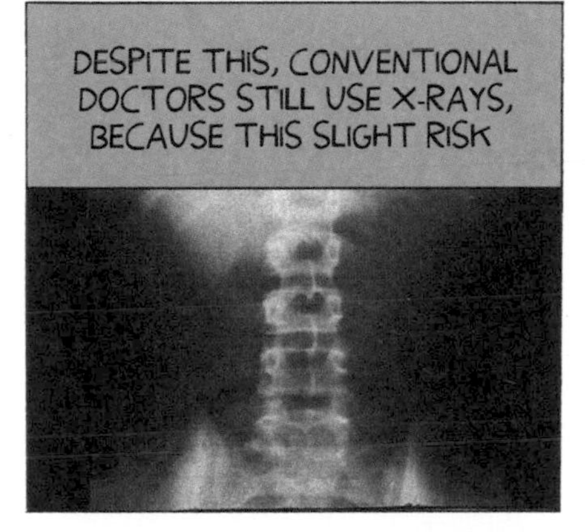

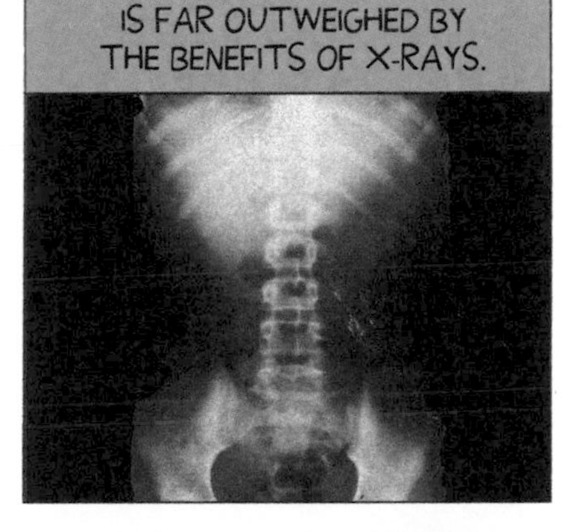

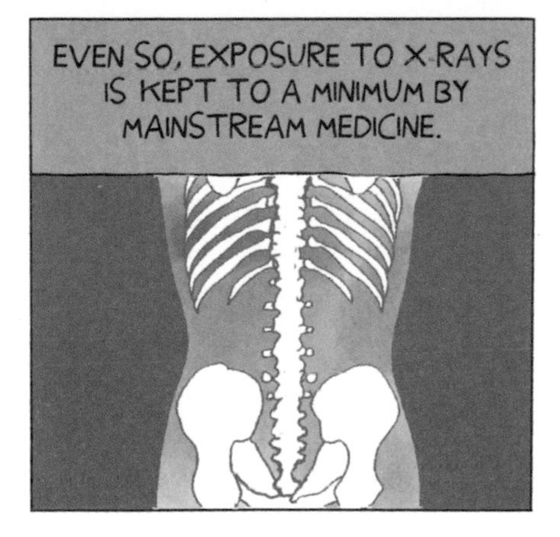

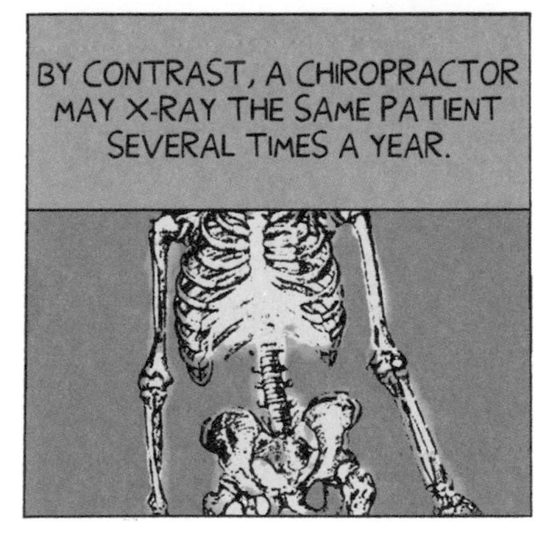

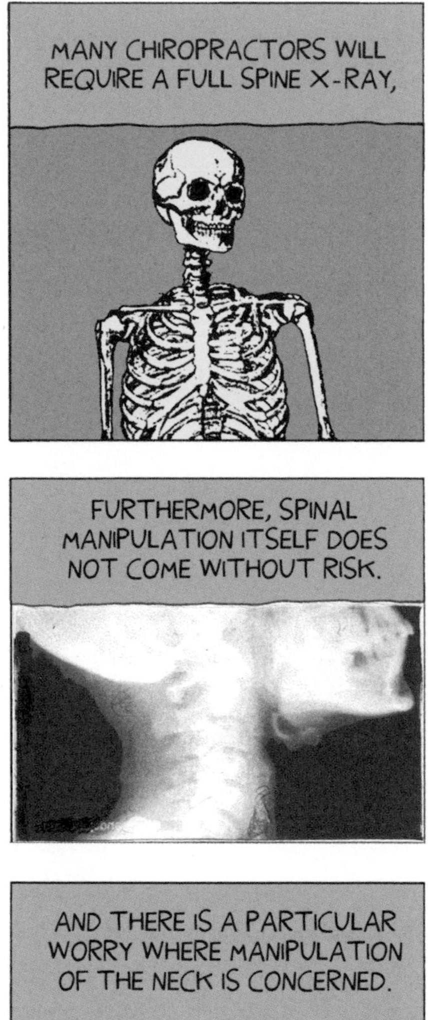
MANY CHIROPRACTORS WILL REQUIRE A FULL SPINE X-RAY,
FURTHERMORE, SPINAL MANIPULATION ITSELF DOES NOT COME WITHOUT RISK.

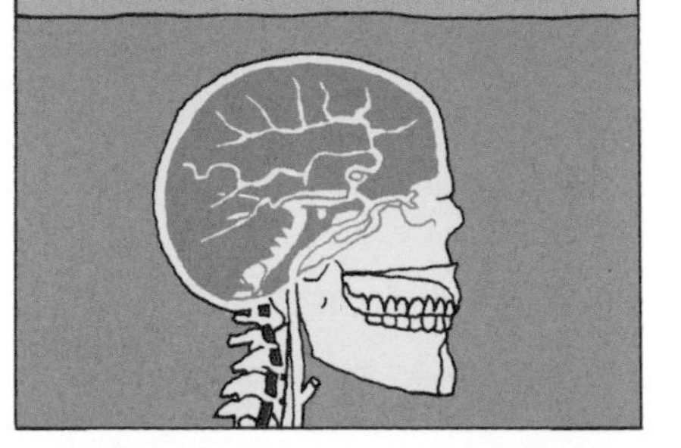
A PROCEDURE THAT DELIVERS A FAR HIGHER DOSE OF RADIATION THAN OTHER X-RAY PRACTICES.

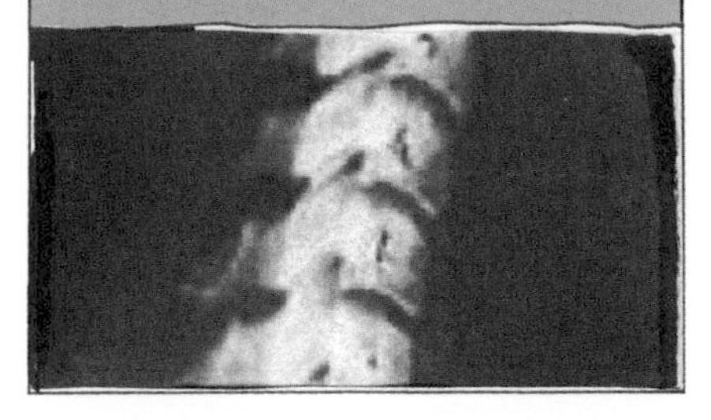
DISLOCATIONS AND FRACTURES AMONG CHIROPRACTIC PATIENTS ARE NOT UNKNOWN.

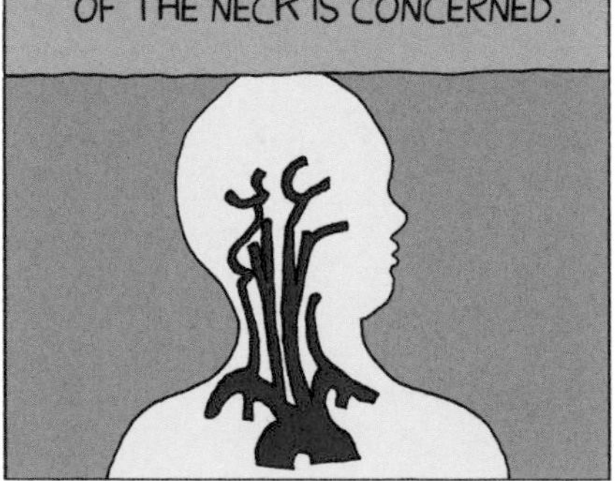
AND THERE IS A PARTICULAR WORRY WHERE MANIPULATION OF THE NECK IS CONCERNED.

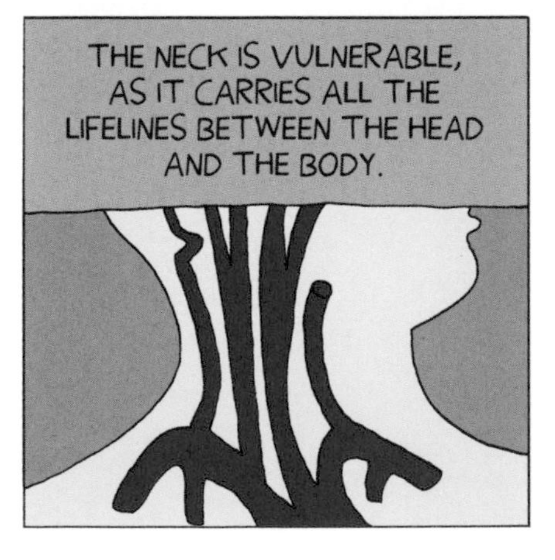
THE NECK IS VULNERABLE, AS IT CARRIES ALL THE LIFELINES BETWEEN THE HEAD AND THE BODY.

DAMAGE TO ARTERIES IN THE NECK CAN LEAD TO A STROKE OR EVEN DEATH.
THESE ARTERIES ARE DEEPLY INTERTWINED INTO THE CERVICAL SECTION OF THE SPINE.
BECAUSE THERE'S OFTEN A DELAY BETWEEN THE INITIAL DAMAGE AND THE RESULTING STROKE
THE CONNECTION BETWEEN CHIROPRACTIC THERAPY AND STROKE WASN'T NOTICED FOR MANY YEARS.
IN RECENT TIMES, HOWEVER, MANY EXAMPLES HAVE COME TO LIGHT:
KRISTI BEDENBAUGH, 24, OF LITTLE MOUNTAIN, SOUTH CAROLINA.

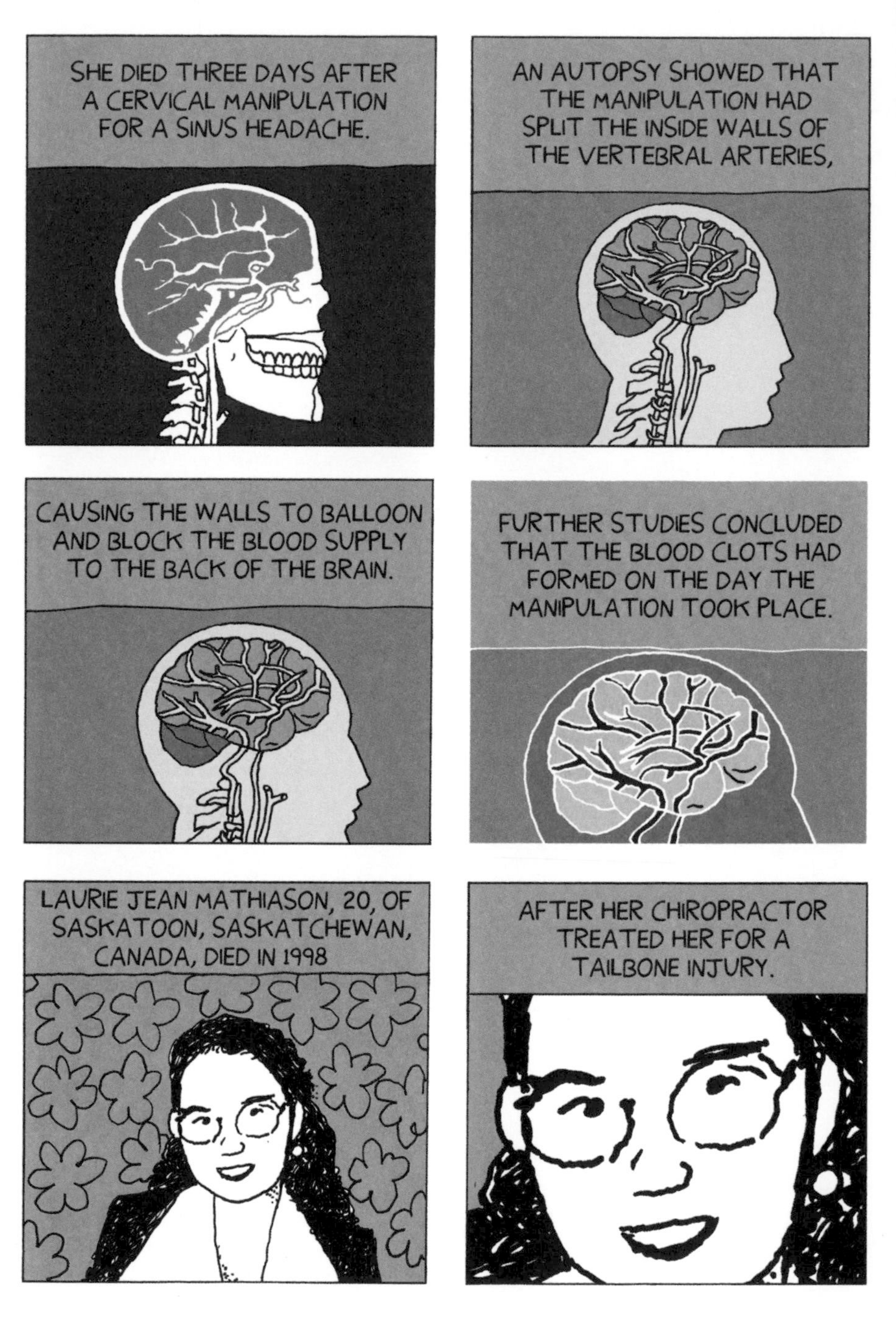
SHE DIED THREE DAYS AFTER A CERVICAL MANIPULATION FOR A SINUS HEADACHE.
AN AUTOPSY SHOWED THAT THE MANIPULATION HAD SPLIT THE INSIDE WALLS OF THE VERTEBRAL ARTERIES,
CAUSING THE WALLS TO BALLOON AND BLOCK THE BLOOD SUPPLY TO THE BACK OF THE BRAIN.
FURTHER STUDIES CONCLUDED THAT THE BLOOD CLOTS HAD FORMED ON THE DAY THE MANIPULATION TOOK PLACE.
LAURIE JEAN MATHIASON, 20, OF SASKATOON, SASKATCHEWAN, CANADA, DIED IN 1998
AFTER HER CHIROPRACTOR TREATED HER FOR A TAILBONE INJURY.

LAURIE'S MOTHER, SHARON MATHIASON.

THE TWIST WAS SO VIOLENT THAT IT TORE HER ARTERY CLEAN THROUGH.

AT THE HOSPITAL WE WERE BOMBARDED WITH DOCTORS COMING INTO THE WAITING ROOM, SAYING,

DON'T YOU KNOW THAT IF YOU GO TO A CHIROPRACTOR, DON'T LET THEM TOUCH YOU ABOVE THE NECK.

BRITTMARIE HARWE, 40, OF WETHERSFIELD, CONNECTICUT.

SHE RECEIVED AN OUT-OF-COURT SETTLEMENT OF $900,000 AFTER A 1993 MANIPULATION PARALYZED ONE OF HER VOCAL CORDS.

THE FEW STUDIES THAT HAVE BEEN DONE SUGGEST THAT THE RISK IS SMALL, BUT REAL.

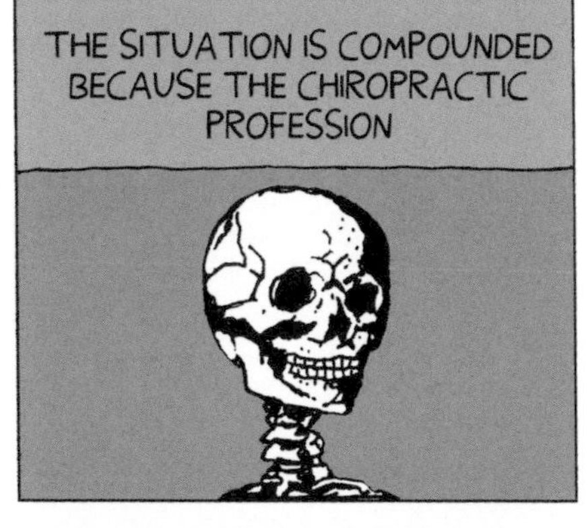

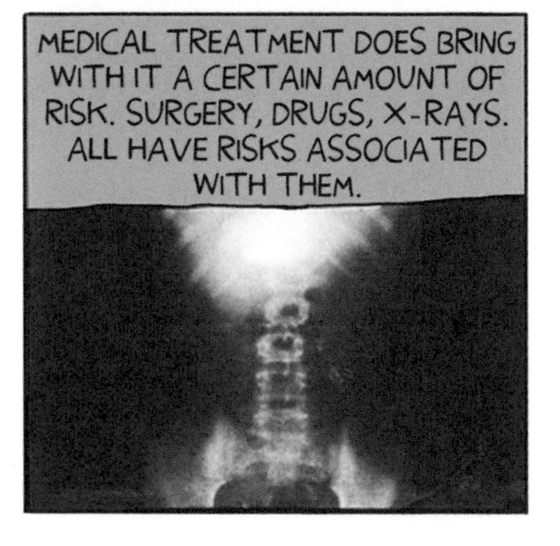

THESE RISKS HAVE TO BE OUTWEIGHED BY THE BENEFITS OF THE TREATMENT IN ORDER FOR THEM TO BE CONSIDERED WORTHWHILE.

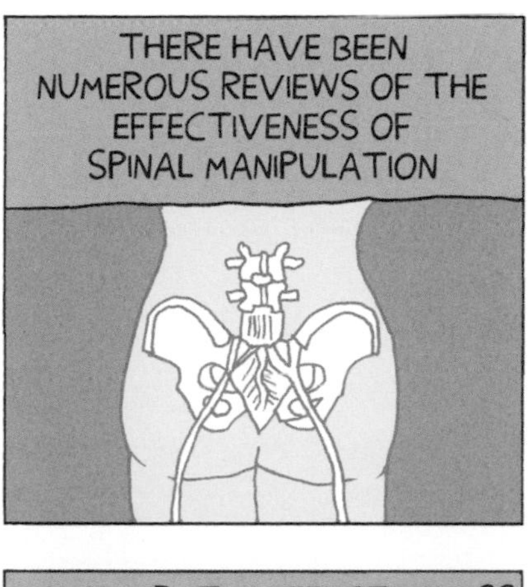
THERE HAVE BEEN NUMEROUS REVIEWS OF THE EFFECTIVENESS OF SPINAL MANIPULATION

ON LOWER BACK PAIN. THESE REVIEWS SUGGEST THAT SPINAL MANIPULATION CAN BRING SOME BENEFIT TO SUFFERERS.
ARRGH!

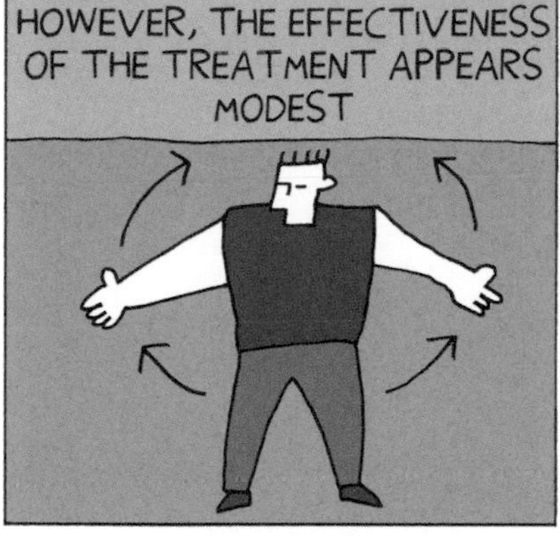
HOWEVER, THE EFFECTIVENESS OF THE TREATMENT APPEARS MODEST

AND IS CERTAINLY NO BETTER THAN TREATMENT GIVEN BY CONVENTIONAL MEDICINE

WHERE DOCTORS MAY RECOMMEND PHYSIOTHERAPY, EXERCISE, OR PRESCRIBE ANTI-INFLAMMATORY DRUGS.

BACK PAIN REMAINS AN EXTREMELY DIFFICULT CONDITION TO TREAT.

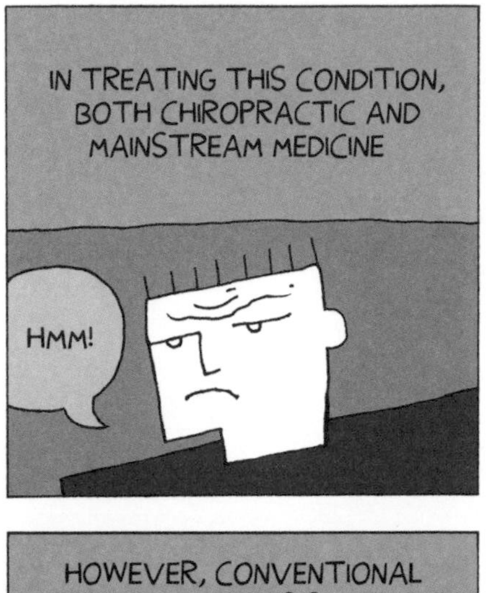
IN TREATING THIS CONDITION, BOTH CHIROPRACTIC AND MAINSTREAM MEDICINE
HMM!

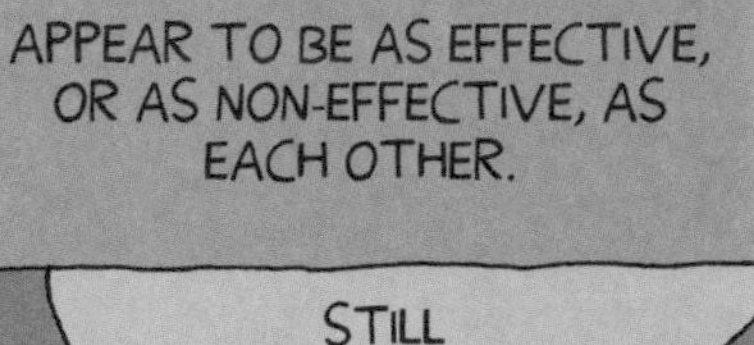
APPEAR TO BE AS EFFECTIVE, OR AS NON-EFFECTIVE, AS EACH OTHER.

STILL HURTS.

HOWEVER, CONVENTIONAL TREATMENT HAS SEVERAL ADVANTAGES OVER CHIROPRACTIC THERAPY.

CHIROPRACTIC THERAPY CAN BE EXPENSIVE.
SO HOW MUCH IS IT, THEN?

BILLS WILL INEVITABLY MOUNT UP OVER REPEATED SESSIONS.
EIGHT HUNDRED DOLLARS, PLEASE.
GAK!

BY COMPARISON, EXERCISE IS FREE AND PAINKILLERS ARE CHEAP.

OTHER PROFESSIONS EXIST THAT OFFER EQUIVALENT SERVICES
SUCH AS PHYSIOTHERAPY,
WHERE YOU WON'T BE GIVEN UNSCIENTIFIC ADVICE.
VACCINATIONS ARE QUITE UNNECESSARY FOR CHILDREN.
WHILE I'VE BEEN DRAWING THIS CHAPTER, I'VE HEARD FROM A FEW PEOPLE WHO
HAVE TOLD ME THAT CHIROPRACTIC THERAPY HAS EASED OR EVEN CURED THEIR BACK PAIN.
TO WHICH I WOULD SAY THAT THEIR SUBJECTIVE EXPERIENCE, HOWEVER POSITIVE,

DOES NOT TRUMP THE WHOLE OF SCIENCE.
THE BENEFITS OF CHIROPRACTIC THERAPY REMAIN SLIGHT FOR THE MAJORITY OF PEOPLE.
CITING A FEW EXAMPLES OF CHIROPRACTIC SUCCESS DOES NOT VALIDATE THIS FORM OF THERAPY.
A REALLY EFFECTIVE CURE FOR BACK PAIN HAS YET TO BE FOUND.
IF A CURE IS EVER DISCOVERED, IT WILL BE FOUND BY SCIENCE-BASED MEDICINE...
AND NOT FROM AN UNSCIENTIFIC AND MYSTICAL DISCIPLINE LIKE CHIROPRACTIC THERAPY.
END

THE MMR VACCINATION SCANDAL

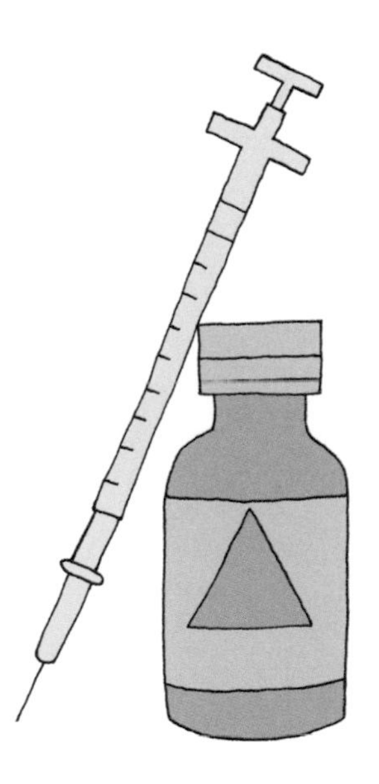

IT'S THE TWENTY-FIRST CENTURY.
AND WE'RE A LONG WAY FROM THE PRE-ENLIGHTENMENT MIDDLE AGES.
THE WORLD HAS BEEN TRANSFORMED BY SCIENTIFIC KNOWLEDGE.
YET SUSPICION OF SCIENCE SEEMS NEVER TO HAVE BEEN HIGHER.
FEAR AND ANGER HAVE OBLITERATED RATIONAL DISCOURSE.

FACTS AND EVIDENCE ARE SEEN AS JUST A MATTER OF OPINION, RATHER THAN A PROVEN TRUTH.
AND BLIND UNREASONING BELIEF IS CONSIDERED AS VALID AS CRITICAL THINKING.
NO MMR
WE'RE WITH WAKEFIELD
MMR. A JAB TOO FAR
THIS IS ANDREW WAKEFIELD,
A BRITISH FORMER SURGEON BEST KNOWN FOR HIS WORK REGARDING THE MEASLES, MUMPS, AND RUBELLA VACCINE (MMR)
AND THE CLAIMED CONNECTION WITH AUTISM AND INFLAMMATORY BOWEL DISEASE.
WAKEFIELD WAS THE LEAD AUTHOR IN A 1998 PAPER PUBLISHED IN THE MEDICAL JOURNAL THE LANCET.
THE LANCET

THE PAPER REPORTED A STUDY OF TWELVE CHILDREN ALL DIAGNOSED WITH AUTISM
IN WHICH THE AUTHORS SUGGESTED A LINK WITH THE MMR VACCINE.
DURING A PRESS CONFERENCE, WAKEFIELD STATED THAT GIVING CHILDREN THE VACCINE IN THREE SEPARATE DOSES
WOULD BE SAFER THAN A SINGLE VACCINATION.
WAKEFIELD'S STUDY WAS NOT SUPPORTED BY THE PAPER, AND SUBSEQUENT PEER REVIEW STUDIES
HAVE NOT SHOWN ANY ASSOCIATION BETWEEN THE MMR VACCINE AND AUTISM.

NONTHELESS, A GLOBAL HEALTH SCARE BEGAN.
FEAR SPREAD AMONG PARENTS WHO WERE UNSURE WHAT IMMUNIZATION CHOICES TO MAKE.
AND PARENTS OF AUTISTIC CHILDREN BEGAN TO QUESTION THE MMR VACCINATION.
HA! HA!
HE WAS FINE BEFORE THAT AWFUL JAB.
BY 2009, HEALTH ORGANIZATIONS IN THE UNITED KINGDOM, THE UNITED STATES, AUSTRALIA, AND MANY OTHER COUNTRIES WERE REPORTING OUTBREAKS OF MEASLES.

MEASLES.
THE UNITED STATES' NATIONAL ACADEMY OF SCIENCES AND THE UK'S NATIONAL HEALTH SERVICE BOTH CONCLUDED
THAT THERE WAS NO EVIDENCE OF A LINK BETWEEN THE MMR VACCINE AND AUTISM.
WHEN I FIRST HEARD ABOUT DR. WAKEFIELD
I THOUGHT, WELL HE'S A POOR SCIENTIST IF HE CAN DRAW SUCH HUGE CONCLUSIONS FROM SUCH A SMALL STUDY.
AND THERE WAS SOMETHING CLEARLY AMISS IN THE FACT HIS RESULTS HAD FAILED TO BE REPLICATED BY ANYONE ELSE.

IT NEVER OCCURRED TO ME THAT WAKEFIELD MIGHT BE A MAN WHO WAS BENEFITING FINANCIALLY FROM THE POISONED ATMOSPHERE
OF FEAR, GUILT AND INFECTIOUS DISEASE THAT HE, IN PART, HAD CREATED.
BUT SUCH WAS THE CASE, AND BRITISH JOURNALIST BRIAN DEER REPORTED THE CONNECTION,
4
FIRST IN THE SUNDAY TIMES (LONDON) AND THEN IN A T.V. DOCUMENTARY ON BRITAIN'S CHANNEL FOUR.
4
TWO YEARS BEFORE THE LANCET PAPER WAS PUBLISHED, DR. WAKEFIELD WAS HIRED BY A LAWYER
RICHARD BARR, A SOLICITOR IN THE UK.

BARR HOPED TO RAISE A CLASS ACTION LAWSUIT AGAINST MMR MANUFACTURERS.
AND WAKEFIELD WAS CONTRACTED TO CONDUCT SCIENTIFIC RESEARCH FOR HIM.
THE AIM WAS TO FIND EVIDENCE THAT THE MMR VACCINE DID HARM.
THIS EVIDENCE WAS INTENDED TO FEATURE IN LITIGATION ON BEHALF OF 1,600 FAMILIES.
FOR THIS WORK, THE DOCTOR WAS PAID $232 AN HOUR,
WHICH WAS PAID OUT OF THE UK'S LEGAL AID FUND.
LEGAL AID
WORKING TOWARD JUSTICE FOR ALL

THE FULL AMOUNT EVENTUALLY CAME TO $675,000 PLUS EXPENSES.
THIS MONEY WAS NEVER DECLARED IN THE LANCET AS IT SHOULD HAVE BEEN.
IT GETS WORSE.
NEARLY NINE MONTHS BEFORE THE PRESS CONFERENCE IN WHICH WAKEFIELD CALLED FOR SINGLE VACCINES,
MMR IS POISON
WE LOVE WAKEFIELD
HE HAD FILED A PATENT FOR HIS OWN SINGLE MEASLES VACCINE.

THIS NEW VACCINE ONLY STOOD A CHANCE OF SUCCESS IF CONFIDENCE IN MMR WAS DAMAGED.
WAKEFIELD'S THEORY WAS THAT BOTH INFLAMMATORY BOWEL DISEASE AND AUTISM
WERE CAUSED BY THE MEASLES (FOUND LIVE AS A NORMAL PART OF THE MMR VACCINE).
BRIAN DEER, THE JOURNALIST, DUG DEEP INTO WAKEFIELD'S FINDINGS.
HE DISCOVERED THAT THE CLINICIANS AND PATHOLOGY SERVICE AT THE ROYAL FREE HOSPITAL,
WHERE WAKEFIELD WORKED, HAD FOUND NOTHING TO IMPLICATE MMR

AND THAT A CLEAR MISMATCH EXISTED BETWEEN WAKEFIELD'S PUBLISHED PAPER
AND THE NATIONAL HEALTH SERVICE RECORDS OF THE CHILDREN IN QUESTION.
DEER CLAIMED THAT WAKEFIELD HAD MANIPULATED DATA RELATING TO THE TWELVE CHILDREN
f behavioural problems had been linked,
had an early adverse reaction to immunisation (rash,
from exposure to first behavioural symptoms was 6·3 days (range
ms because children were not toilet trained at the time or because
eceived monovalent measles vaccine at 15 months, after which his
n was made with the vaccine at this time. He received a dose of
other described a striking deterioration in his behaviour that she di
bella vaccine at 16 months. At 18 months he developed recurrent
nterest in his sibling and lack of play.
TO LOOK AS IF A NEW SYNDROME HAD BEEN DISCOVERED.
monovalent measles vaccine
de with the vaccine at this ti
ibed a striking deterioration i
e at 16 months. At 18 month
sibling and lack of play.

THE GENERAL MEDICAL COUNCIL (GMC) AND THE LANCET TOOK ACTION.
MMR NEWS

THE LANCET RETRACTED WAKEFIELD'S PAPER.

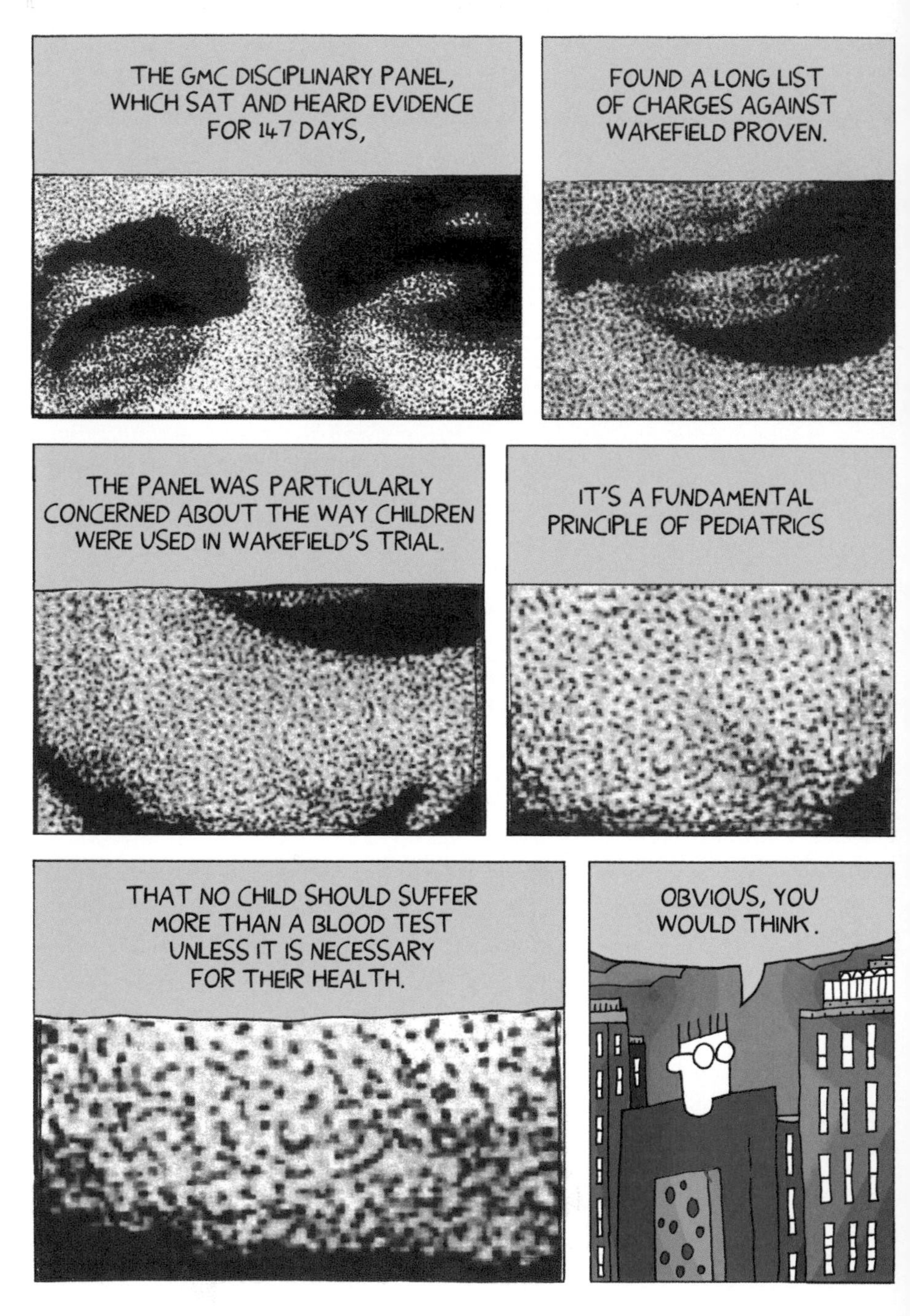
THE GMC DISCIPLINARY PANEL, WHICH SAT AND HEARD EVIDENCE FOR 147 DAYS,
FOUND A LONG LIST OF CHARGES AGAINST WAKEFIELD PROVEN.
THE PANEL WAS PARTICULARLY CONCERNED ABOUT THE WAY CHILDREN WERE USED IN WAKEFIELD'S TRIAL.
IT'S A FUNDAMENTAL PRINCIPLE OF PEDIATRICS
THAT NO CHILD SHOULD SUFFER MORE THAN A BLOOD TEST UNLESS IT IS NECESSARY FOR THEIR HEALTH.
OBVIOUS, YOU WOULD THINK.

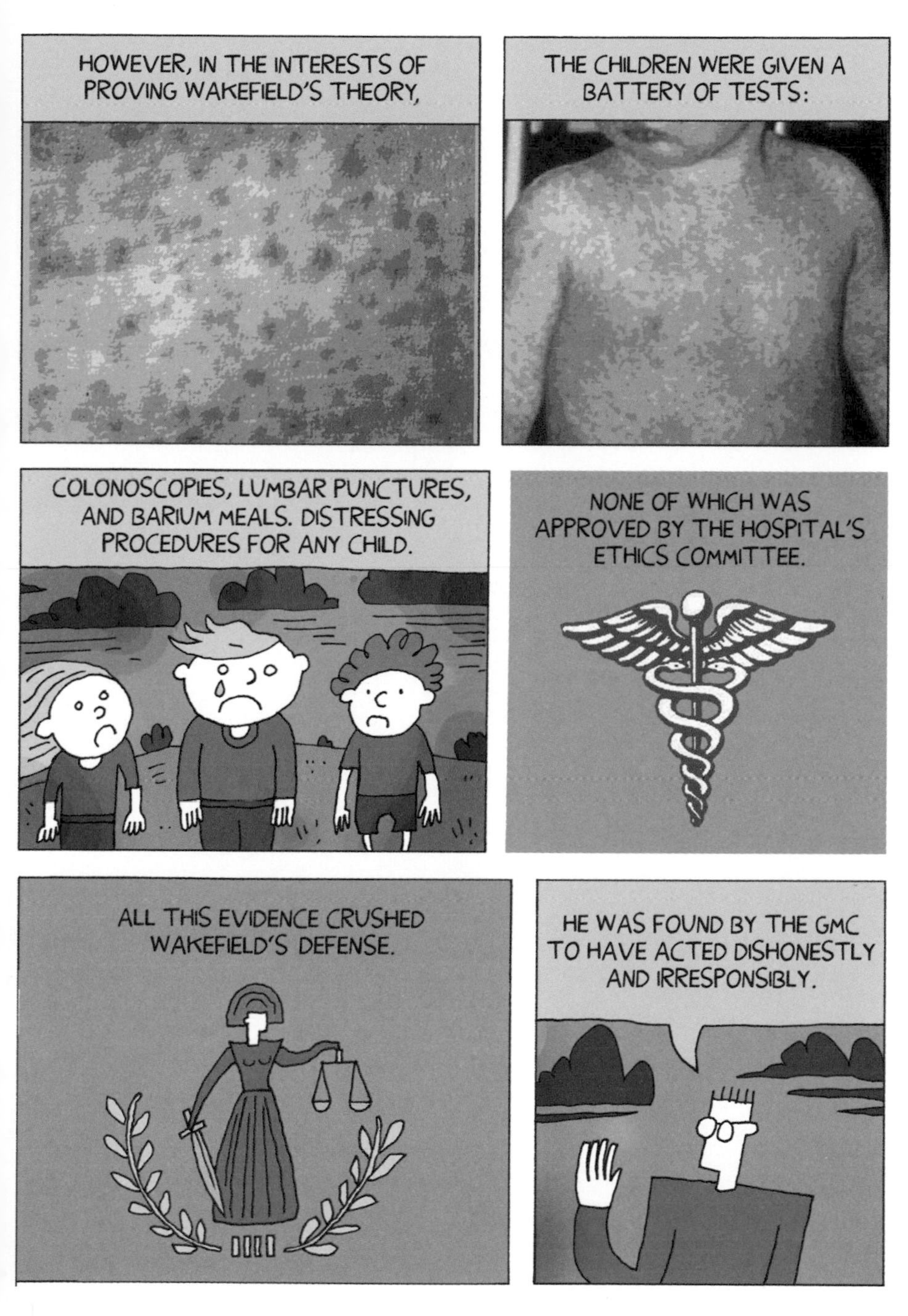
HOWEVER, IN THE INTERESTS OF PROVING WAKEFIELD'S THEORY,
THE CHILDREN WERE GIVEN A BATTERY OF TESTS:
COLONOSCOPIES, LUMBAR PUNCTURES, AND BARIUM MEALS. DISTRESSING PROCEDURES FOR ANY CHILD.
NONE OF WHICH WAS APPROVED BY THE HOSPITAL'S ETHICS COMMITTEE.
ALL THIS EVIDENCE CRUSHED WAKEFIELD'S DEFENSE.
HE WAS FOUND BY THE GMC TO HAVE ACTED DISHONESTLY AND IRRESPONSIBLY.

HE LIVES IN THE UNITED STATES, WHERE HE'S SEEN AS A HERO BY THE ANTI-VACCINE MOVEMENT.

ENDORSED BY CELEBRITIES SUCH AS JIM CARREY

BY VACCINE MANUFACTURERS AND THEIR PAID LACKEYS IN THE MEDIA AND SCIENTIFIC ESTABLISHMENT.

FOR THE TEN YEARS THAT THE MMR CONTROVERSY RAGED
THE MEDIA IN BRITAIN FELL OVER THEMSELVES PROMOTING AND SUPPORTING DR. WAKEFIELD.
BLAH! BLAH!
MOST NEWSPAPERS, WITH ONLY A FEW EXCEPTIONS, UNCRITICALLY SWALLOWED WHOLE HIS STUDY'S FEEBLE EVIDENCE.
SENSATIONALLY HIGHLIGHTING THE STORY
WHILE DOWNPLAYING AND EVEN IGNORING STUDIES WHICH FOUND NO CONNECTION BETWEEN MMR AND AUTISM.
WAKEFIELD WAS FAR FROM ALONE IN CREATING THIS HEALTH SCARE.

WHERE SCIENCE IS CONCERNED, JOURNALISTS LIKE BRIAN DEER ARE RARE.
Interview
ONE IN THE ARM FOR MMR
THE SU
Revealed: M
INSTEAD WE HAVE POOR AND BIASED SCIENCE REPORTING FOR SENSATION-HUNGRY NEWSPAPERS.
DAILY SAME
BAN THIS SICK FILTH
WHO SPREAD ILL-INFORMED OPINIONS LIKE THEY WERE SOLID GOLD TRUTHS.
DAILEY SP
IN THE TWENTY-FIRST CENTURY, IS IT TOO MUCH TO ASK JOURNALISTS TO DO BASIC FACT-CHECKING?
DAILEY SP
AND THAT EDITORS WOULD ASSIGN SCIENCE STORIES TO REPORTERS WHO HAVE A KNOWLEDGE OF THE SUBJECT.
DAILEY S
IS THAT TOO MUCH TO ASK?
END

EVOLUTION

HEY, EXPLAIN THIS! WHY ARE THERE STILL APES IF THEY WERE ALLEGEDLY OUR ANCESTORS?

THAT'S A VERY GOOD QUESTION AND ONE THAT DESERVES AN ANSWER.

THE EXPLANATION IS THAT APES AND HUMANS ARE ON A BRANCH OF THE EVOLUTIONARY TREE
THAT EVOLVED FROM AN EARLIER COMMON ANCESTOR.
BONOBOS
CHIMPANZEES
HUMANS
SPLIT FIVE MILLION YEARS AGO.
THESE SPECIES ARE CONNECTED, NOT FROM ONE TO ANOTHER IN THE PRESENT TIME, BUT IN THE DEEP PAST. IT IS THIS EARLIER COMMON ANCESTOR THAT IS LONG EXTINCT.
GIBBONS
ORANGUTANS
GORILLAS
BONOBOS
CHIMPANZEES
HUMANS

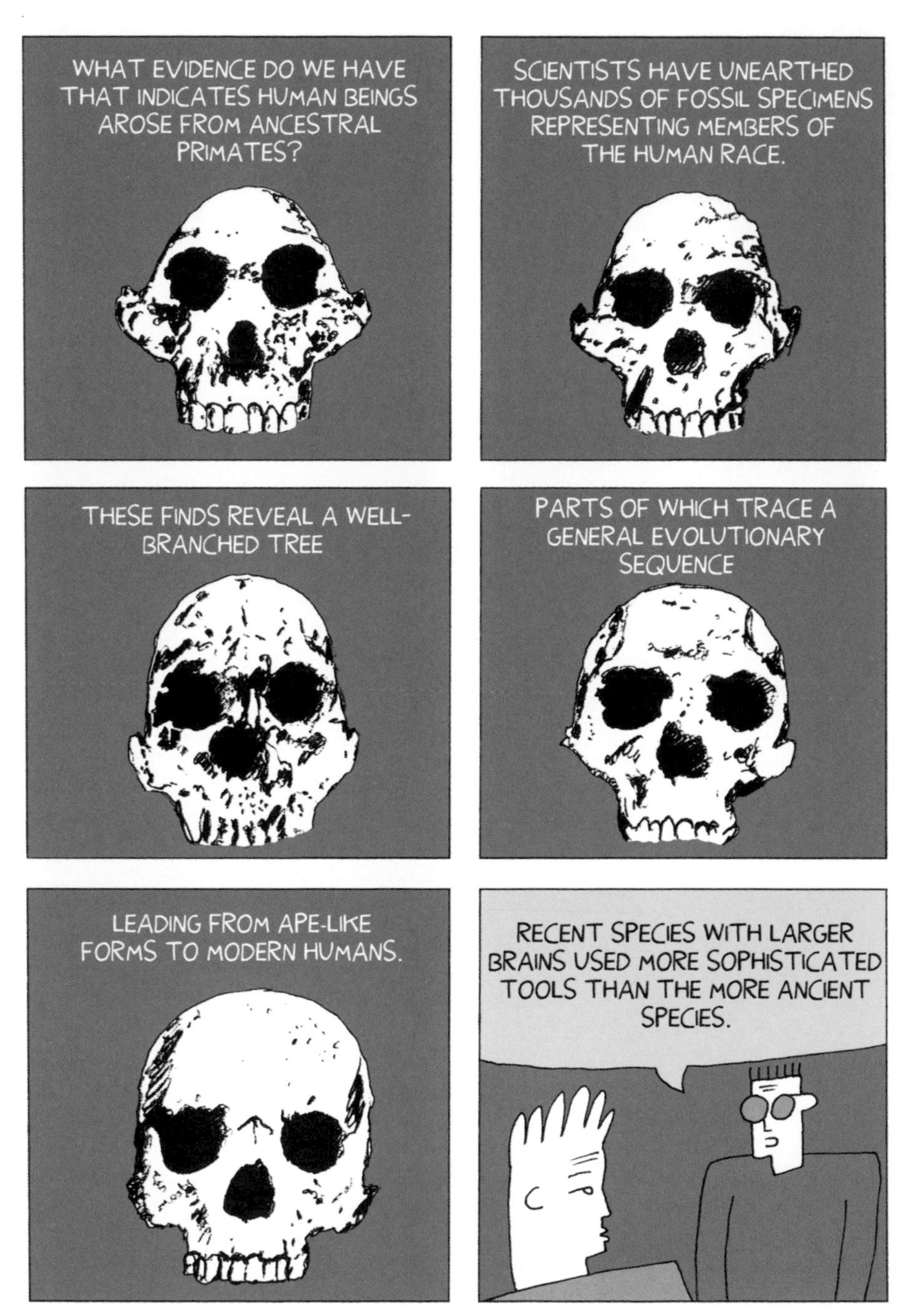
WHAT EVIDENCE DO WE HAVE THAT INDICATES HUMAN BEINGS AROSE FROM ANCESTRAL PRIMATES?
SCIENTISTS HAVE UNEARTHED THOUSANDS OF FOSSIL SPECIMENS REPRESENTING MEMBERS OF THE HUMAN RACE.
THESE FINDS REVEAL A WELL-BRANCHED TREE
PARTS OF WHICH TRACE A GENERAL EVOLUTIONARY SEQUENCE
LEADING FROM APE-LIKE FORMS TO MODERN HUMANS.
RECENT SPECIES WITH LARGER BRAINS USED MORE SOPHISTICATED TOOLS THAN THE MORE ANCIENT SPECIES.

MOLECULAR BIOLOGY HAS PROVIDED STRONG EVIDENCE OF THE CLOSE RELATIONSHIP BETWEEN HUMANS AND OTHERS IN THE APE FAMILY.
OOK!
ANALYSIS OF PROTEINS AND GENES HAS SHOWN THAT HUMANS ARE GENETICALLY SIMILAR
TO CHIMPANZEES AND GORILLAS AND LESS SIMILAR TO ORANGUTANS AND OTHER PRIMATES.
BUT ALL THIS IS JUST A THEORY, ISN'T IT?
A GOOD THEORY MAKES PREDICTIONS ABOUT WHAT WE WOULD FIND IF WE LOOK CLOSELY AT NATURE.
AND, IF THESE PREDICTIONS ARE MET, IT GIVES US CONFIDENCE THAT THE THEORY IS TRUE.

HERE ARE SOME EVOLUTIONARY PREDICTIONS.
WE SHOULD BE ABLE TO FIND EVIDENCE FOR EVOLUTIONARY CHANGE IN THE FOSSIL RECORD.
THE DEEPEST AND OLDEST LAYERS OF ROCK WOULD CONTAIN THE FOSSILS OF MORE PRIMITIVE SPECIES.
ORGANISMS RESEMBLING PRESENT-DAY SPECIES SHOULD BE FOUND IN THE MOST RECENT LAYERS OF ROCK.

WE SHOULD BE ABLE TO SEE SOME SPECIES CHANGING OVER TIME, FORMING LINEAGES.

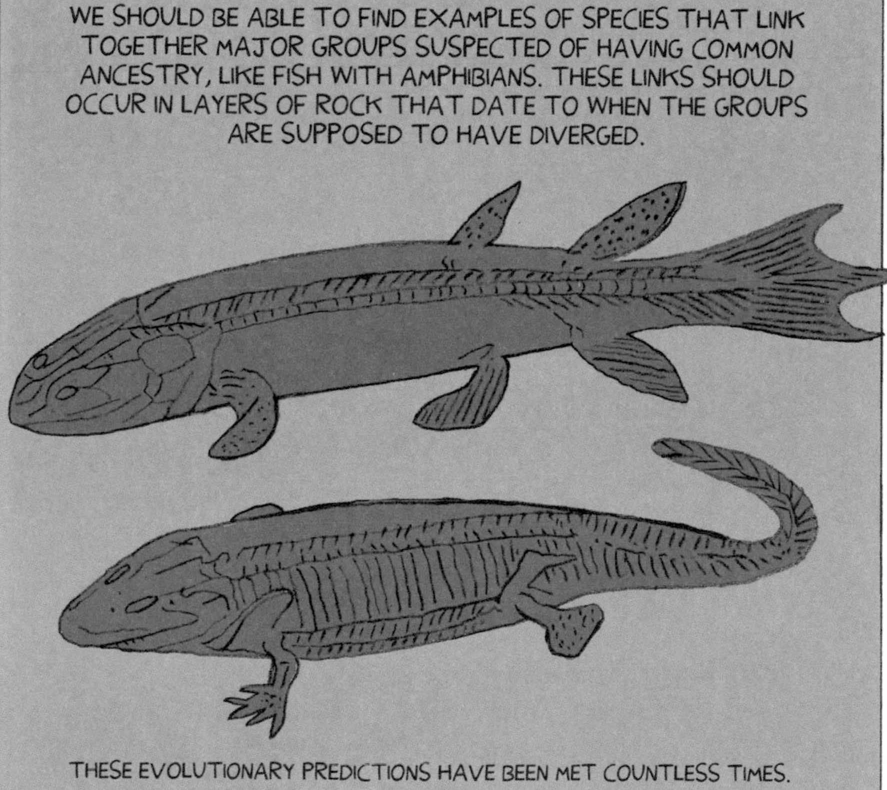
WE SHOULD BE ABLE TO FIND EXAMPLES OF SPECIES THAT LINK TOGETHER MAJOR GROUPS SUSPECTED OF HAVING COMMON ANCESTRY, LIKE FISH WITH AMPHIBIANS. THESE LINKS SHOULD OCCUR IN LAYERS OF ROCK THAT DATE TO WHEN THE GROUPS ARE SUPPOSED TO HAVE DIVERGED.
THESE EVOLUTIONARY PREDICTIONS HAVE BEEN MET COUNTLESS TIMES.

FURTHERMORE, WE SHOULD BE ABLE TO FIND CASES OF IMPERFECT ADAPTATION IN WHICH EVOLUTION
HAS NOT BEEN ABLE TO ACHIEVE THE SAME DEGREE OF EFFICIENCY AS A CREATOR WOULD.
FOR EXAMPLE, IT WOULD BE BETTER FOR MEN IF OUR TESTES FORMED OUTSIDE OUR BODIES
WHERE THE COOLER TEMPERATURE IS BETTER FOR SPERM.
HOWEVER, THE TESTES BEGIN DEVELOPMENT IN THE ABDOMEN. WHEN THE FETUS IS SIX OR SEVEN MONTHS OLD
THEY MIGRATE DOWN INTO THE SCROTUM THROUGH TWO CHANNELS

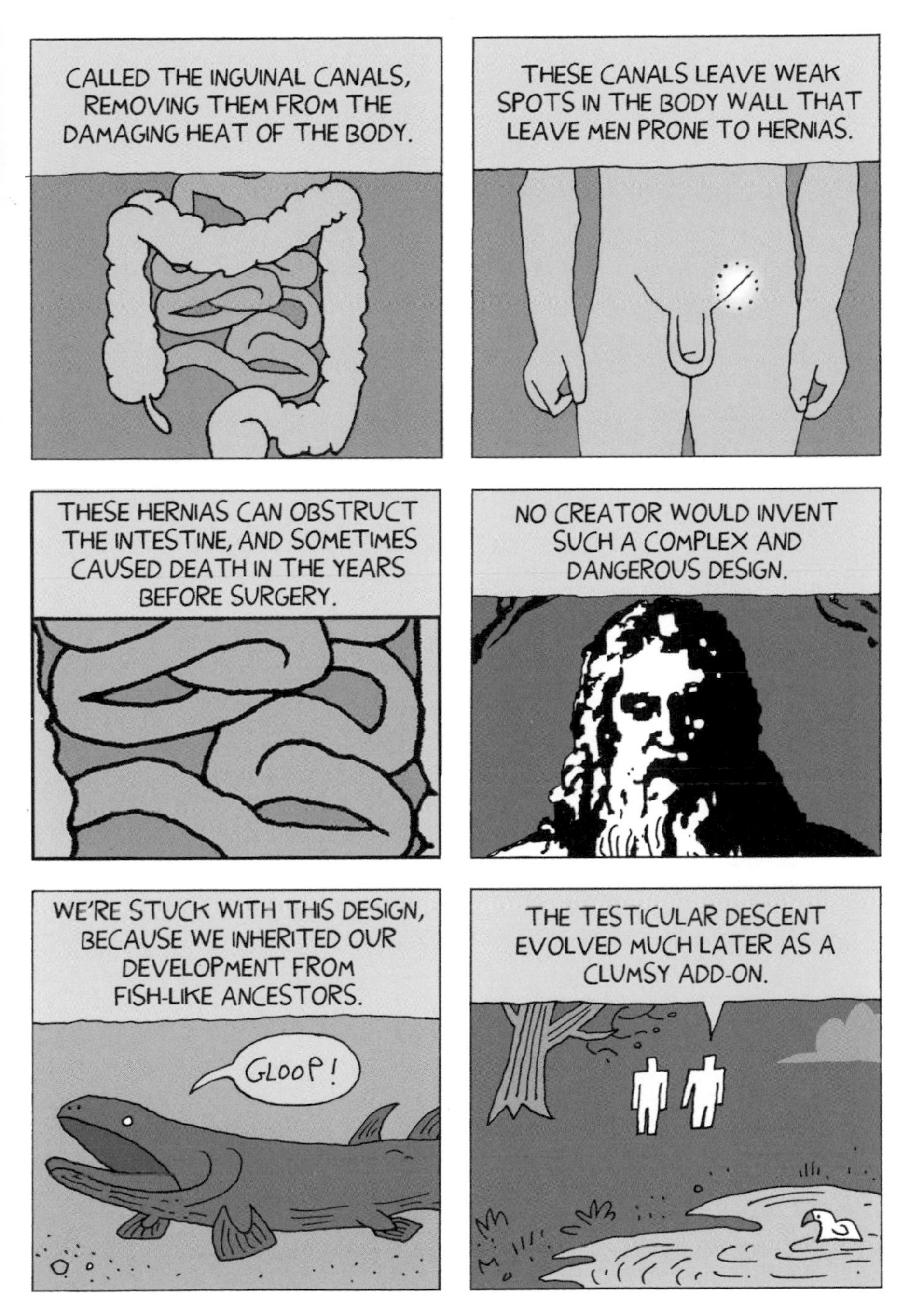
CALLED THE INGUINAL CANALS, REMOVING THEM FROM THE DAMAGING HEAT OF THE BODY.
THESE CANALS LEAVE WEAK SPOTS IN THE BODY WALL THAT LEAVE MEN PRONE TO HERNIAS.
THESE HERNIAS CAN OBSTRUCT THE INTESTINE, AND SOMETIMES CAUSED DEATH IN THE YEARS BEFORE SURGERY.
NO CREATOR WOULD INVENT SUCH A COMPLEX AND DANGEROUS DESIGN.
WE'RE STUCK WITH THIS DESIGN, BECAUSE WE INHERITED OUR DEVELOPMENT FROM FISH-LIKE ANCESTORS.
GLOOP!
THE TESTICULAR DESCENT EVOLVED MUCH LATER AS A CLUMSY ADD-ON.

EACH SPECIES IS BUILT ON OLDER DESIGNS. THE SLATE IS NOT WIPED CLEAN EACH TIME.

NEW PARTS EVOLVE FROM OLD ONES, AND THEY HAVE TO WORK WITH PARTS THAT ARE ALREADY THERE.

THIS CAN LEAD TO SOME STRANGE DEVELOPMENTAL FEATURES.

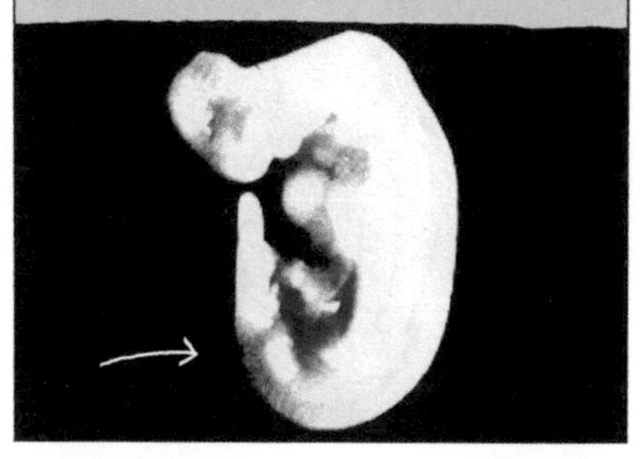
EMBRYONIC WHALES AND DOLPHINS FORM HIND LIMB BUDS.

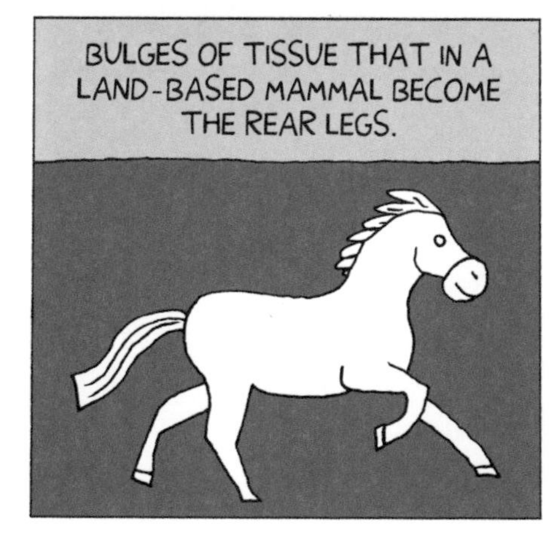
BULGES OF TISSUE THAT IN A LAND-BASED MAMMAL BECOME THE REAR LEGS.

BUT IN THE MARINE MAMMALS THESE BUDS ARE REABSORBED SOON AFTER THEY'RE FORMED.

BALEEN WHALES, THAT LACK TEETH, BUT WHOSE ANCESTORS WERE TOOTHED WHALES...
DEVELOP EMBRYONIC TEETH THAT VANISH BEFORE BIRTH.
BUT WHAT MAKES THINGS EVOLVE? HOW CAN ONE THING BECOME ANOTHER?
LIFE IS NOT A FLUID THAT FLOWS AND RESHAPES ITSELF DEPENDING ON CONTEXT.
WELL, IT KIND OF IS. THAT'S EXACTLY HOW YOU SHOULD LOOK AT IT.
AND IT IS THE PRIMAL FORCES OF TIME AND GENETICS THAT DO THE MOLDING.

CHECK OUT THESE BUGS DOWN HERE.
HALF ARE GRAY AND HALF ARE ORANGE.
THE ORANGE BEETLES ARE MUCH MORE ATTRACTIVE TO PREDATORS
AND THEREFORE HAVE LESS CHANCE TO REPRODUCE THAN DO THE GRAY BEETLES.
THE CAMOUFLAGED GRAY BEETLES PASS THE GENETIC COLOR TRAIT ONTO THEIR OFFSPRING
SO THAT EVENTUALLY ALL THE INDIVIDUALS IN THE POPULATION WILL BE GRAY.

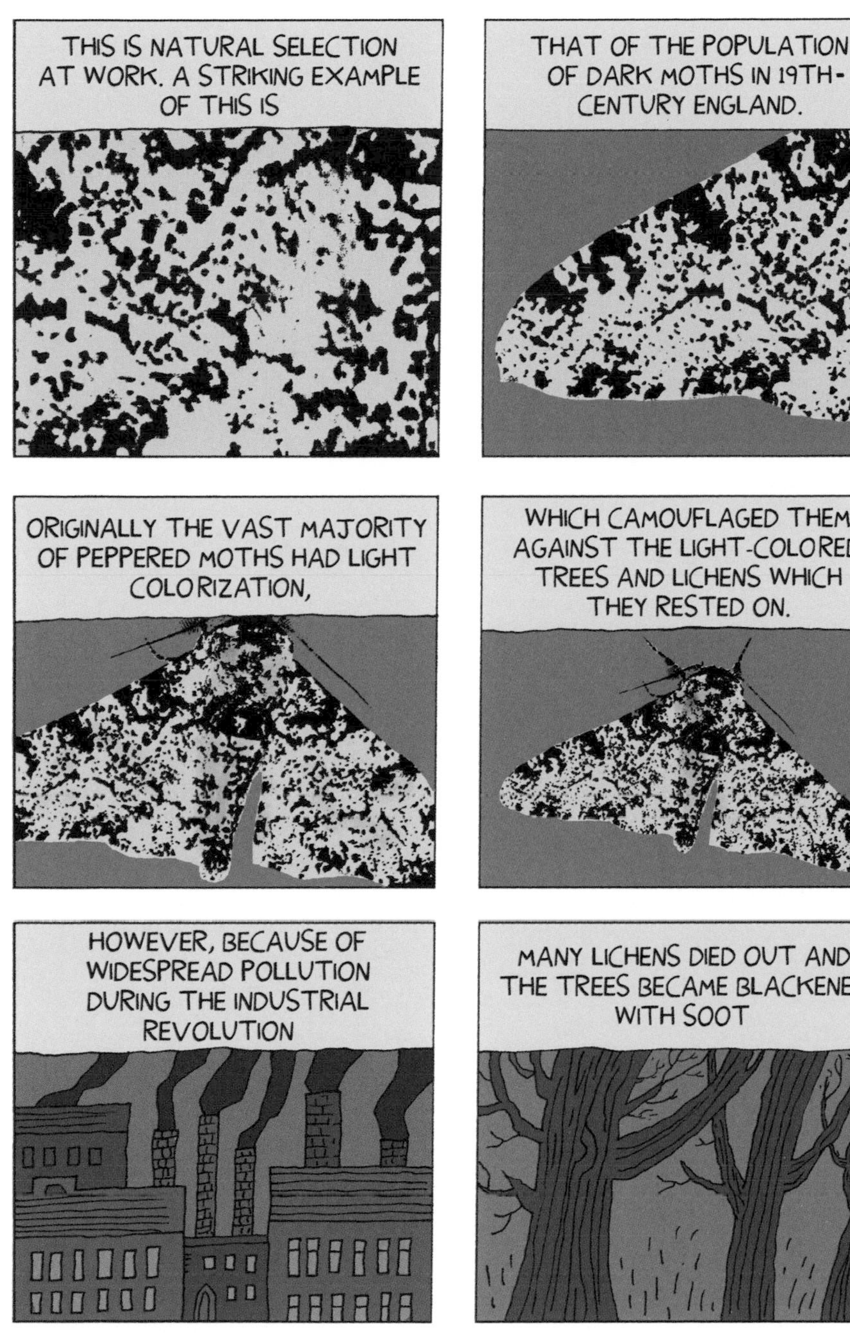
THIS IS NATURAL SELECTION AT WORK. A STRIKING EXAMPLE OF THIS IS
THAT OF THE POPULATION OF DARK MOTHS IN 19TH-CENTURY ENGLAND.
ORIGINALLY THE VAST MAJORITY OF PEPPERED MOTHS HAD LIGHT COLORIZATION,
WHICH CAMOUFLAGED THEM AGAINST THE LIGHT-COLORED TREES AND LICHENS WHICH THEY RESTED ON.
HOWEVER, BECAUSE OF WIDESPREAD POLLUTION DURING THE INDUSTRIAL REVOLUTION
MANY LICHENS DIED OUT AND THE TREES BECAME BLACKENED WITH SOOT

CAUSING MOST OF THE LIGHT-COLORED MOTHS TO DIE OFF FROM PREDATION.
AT THE SAME TIME THE DARK-COLORED MOTHS FLOURISHED.
SINCE THEN THERE HAVE BEEN IMPROVED ENVIRONMENTAL STANDARDS
AND SO LIGHT-COLORED PEPPERED MOTHS AGAIN BECAME COMMON.
OH YEAH! WELL, I HAPPEN TO KNOW THAT THE WHOLE PEPPERED MOTH THING HAS BEEN REVEALED AS A HOAX.
THERE WERE STUDIES OF THIS PHENOMENON BACK IN THE 1950s.

IN MORE RECENT TIMES, BIOLOGISTS HAVE CRITICIZED SOME ASPECTS OF THE METHODOLOGY OF THIS RESEARCH.
AS A RESULT, CERTAIN CRITICS HAVE WRONGLY CLAIMED THAT THE PEPPERED MOTH STORY WAS A HOAX.
KOFF!
BUT THE BASIC ELEMENTS OF THE STORY ARE CORRECT. THE POPULATION OF DARK MOTHS ROSE AND FELL IN PARALLEL WITH INDUSTRIAL POLLUTION.
OF THE SEVERAL FACTORS KNOWN TO PRODUCE EVOLUTIONARY CHANGE,
ONLY NATURAL SELECTION IS CONSISTENT WITH THE PATTERN OF CHANGE SEEN IN THE MOTH POPULATION.
WELL, ALL THAT IS WELL AND GOOD, BUT HOW DO YOU EXPLAIN MORE COMPLEX EVOLUTIONARY PROCESSES?

HOW DO YOU EXPLAIN THE EVOLUTION OF THE EYE?

HOW CAN SOMETHING SO COMPLEX HAVE DEVELOPED THROUGH NATURAL SELECTION, EVEN AFTER MILLIONS OF YEARS?

LET ME EXPLAIN. DIFFERENT TYPES OF EYES HAVE EMERGED IN EVOLUTIONARY HISTORY.

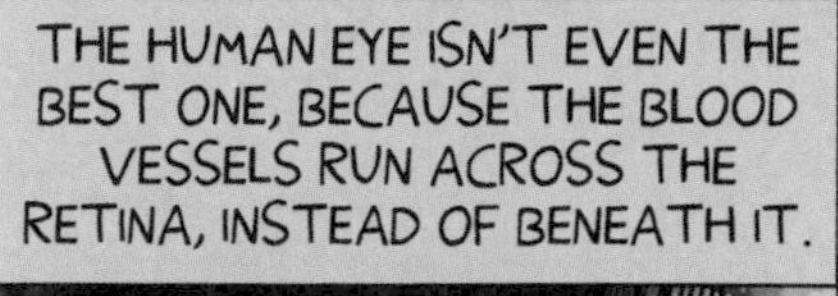
THE HUMAN EYE ISN'T EVEN THE BEST ONE, BECAUSE THE BLOOD VESSELS RUN ACROSS THE RETINA, INSTEAD OF BENEATH IT.

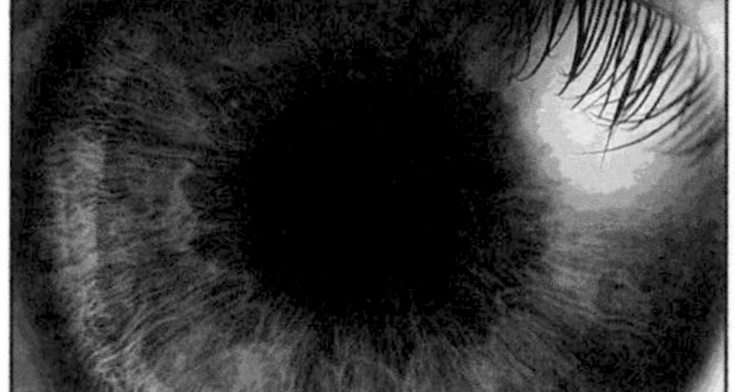

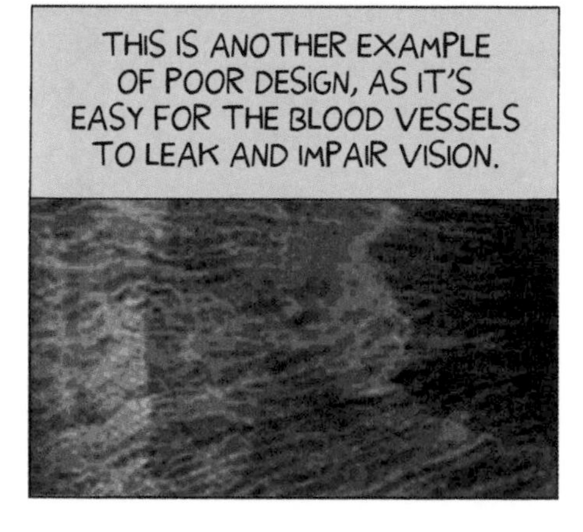
THIS IS ANOTHER EXAMPLE OF POOR DESIGN, AS IT'S EASY FOR THE BLOOD VESSELS TO LEAK AND IMPAIR VISION.

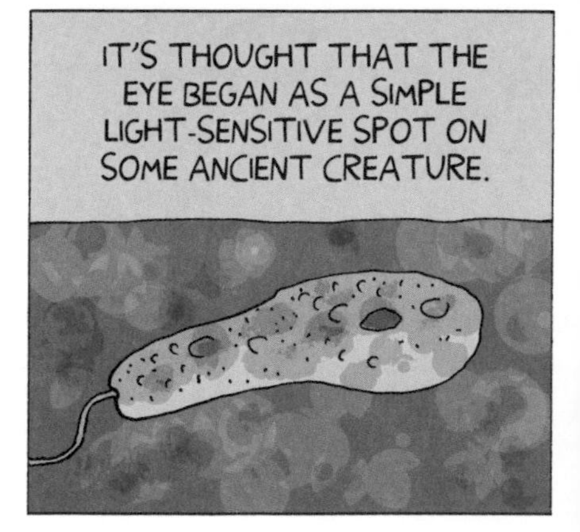
IT'S THOUGHT THAT THE EYE BEGAN AS A SIMPLE LIGHT-SENSITIVE SPOT ON SOME ANCIENT CREATURE.

A SPOT SENSITIVE ENOUGH TO DETECT A PREDATOR'S SHADOW.
PHOTORECEPTOR LAYER
NERVE FIBERS
THIS GAVE THE SIGHTED CREATURE AN ADVANTAGE OVER OTHERS OF ITS TYPE.
ITS COMPETITORS DIED, WHILE IT THRIVED, PASSING ITS LIGHT-SENSITIVE SPOT ON TO ITS DESCENDANTS.
EVENTUALLY A RANDOM MUTATION CREATED A DEPRESSION IN THE LIGHT-SENSITIVE PATCH.
A DEEPENING PIT THAT ENABLED IT TO DETECT THE ANGLE OF LIGHT MORE SHARPLY.
AT EVERY POINT OF ADVANTAGE THE CREATURE PASSED THESE IMPROVEMENTS ONTO ITS OFFSPRING.

AFTER MANY GENERATIONS THE PIT'S OPENING NARROWED, SO THAT LIGHT ENTERED THROUGH A SMALL APERTURE
MUCH LIKE A PINHOLE CAMERA, ALLOWING THE EYE TO PRODUCE A RUDIMENTARY IMAGE.
OPTIC NERVE
OVER TIME, A LENS FORMED AT THE FRONT OF THE EYE AND CLEAR FLUID FILLED THE CHAMBER,
WHILE THE LIGHT-SENSITIVE PATCH BECAME A RETINA.
EYES CORRESPONDING TO EVERY STAGE IN THIS SEQUENCE
HAVE BEEN FOUND IN EXISTING LIVING SPECIES.

BUT WHAT MAKES THESE CHANGES HAPPEN IN THE FIRST PLACE?

FOR THE ANSWER TO THAT WE HAVE TO LOOK DEEP INTO THE MECHANISMS OF LIFE ITSELF.

HOW IT REPLICATES AND THE MISTAKES THAT CAN HAPPEN WHEN IT DOES.

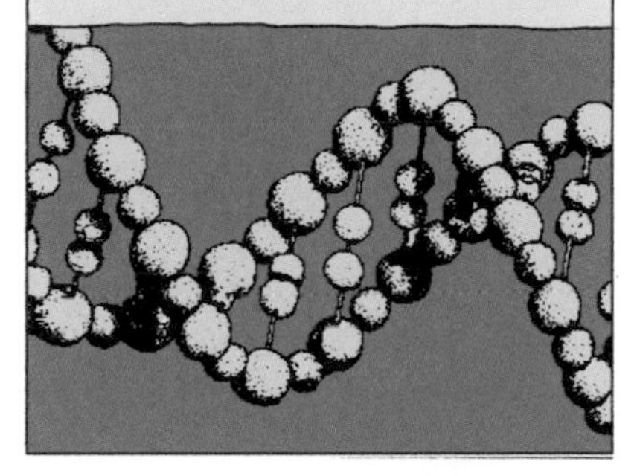
WITHIN EVERY CELL THERE IS A SUBSTANCE CALLED DEOXYRIBONUCLEIC ACID (DNA).

DNA IS VERY FINE AND TIGHTLY COILED. THERE MAY BE AS MUCH AS THREE FEET (1 METER) IN A SINGLE CELL.

DNA IS REALLY A CODE. IT'S DIVIDED UP INTO SECTIONS CALLED GENES.

YOU HAVE TO THINK OF DNA AS A VAST CHEMICAL INFORMATION DATABASE

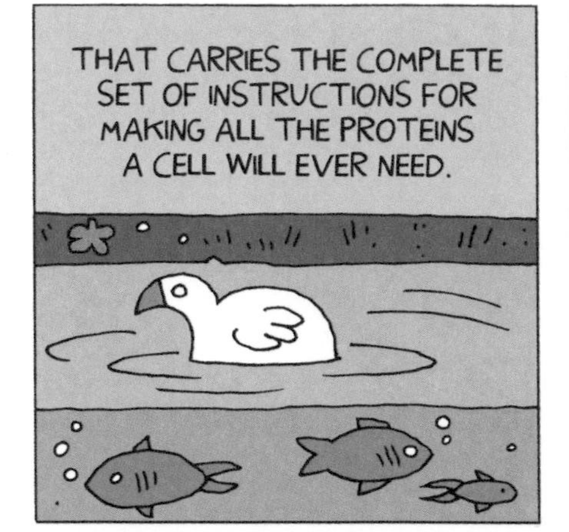
THAT CARRIES THE COMPLETE SET OF INSTRUCTIONS FOR MAKING ALL THE PROTEINS A CELL WILL EVER NEED.

THIS GENETIC INFORMATION IS NOT SO MUCH A BLUEPRINT FOR BUILDING AN ORGANISM

AS A RECIPE, OUT OF WHICH COMPLEX STRUCTURES CAN THEN EMERGE.

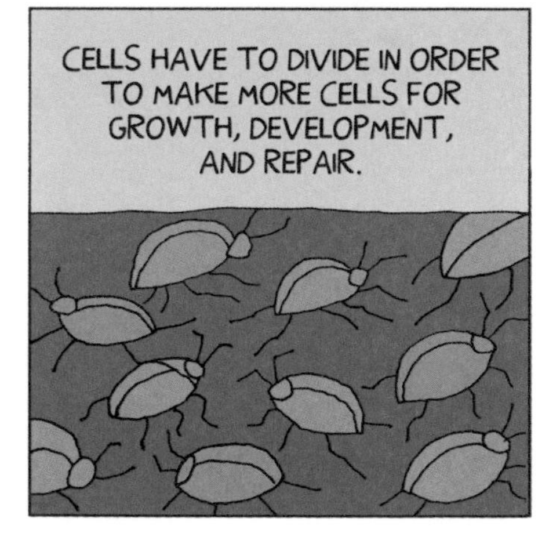
CELLS HAVE TO DIVIDE IN ORDER TO MAKE MORE CELLS FOR GROWTH, DEVELOPMENT, AND REPAIR.

DURING THESE DIVISIONS, MISTAKES IN THE COPYING OF THE DNA STRANDS SOMETIMES HAPPEN.

APART FROM THE COPYING ERRORS THAT OCCUR DURING REPLICATION,
MUTATIONS CAN BE CAUSED BY VIRUSES, RADIATION, AND CHEMICALS.
WHEN MUTATIONS HAPPEN THEY CAN HAVE NO EFFECT, THEY CAN BE DAMAGING,
OR THEY CAN OFFER SOME SLIGHT ADVANTAGE IN THE STRUGGLE FOR LIFE.
LIKE THE APPEARANCE OF THE LIGHT-SENSITIVE SPOT ON THE ANCIENT CREATURE PREVIOUSLY DISCUSSED.
CELL DIVISION IS A RANDOM PROCESS, MUCH LIKE THE SHUFFLING OF CARDS.

SOMETIMES AN ORGANISM CAN BE DEALT A USEFUL HAND, AND THEN BECAUSE OF NATURAL SELECTION THESE USEFUL GENETIC TRAITS BUILD UP OVER TIME,

CREATING THE COMPLEX EVOLUTIONARY CHANGES WE SEE.

THE HUNDREDS OF MILLIONS OF YEARS THAT LIFE HAS EXISTED ON EARTH

HAS BEEN MORE THAN ENOUGH TIME

FOR THESE INCREMENTAL CHANGES TO BUILD UP INTO HUGE EFFECTS.

THESE EVOLUTIONARY PROCESSES HAVEN'T ENDED.

WE SEE THIS IN THE WAY BACTERIA HAVE BECOME RESISTANT TO ANTIBIOTICS.
AND IN SPECIFIC EXAMPLES LIKE THE CHANGE IN THE BEAK SIZE OF THE MEDIUM GROUND FINCH OF THE GALAPAGOS ISLANDS.
IN 1977 A SEVERE DROUGHT REDUCED THE SUPPLY OF SEEDS IN THE GALAPAGOS.
GALAPAGOS ISLANDS
PACIFIC OCEAN
THIS FINCH, WHICH NORMALLY PREFERRED SMALL SOFT SEEDS
WAS FORCED TO TURN TO LARGER AND HARDER SEEDS.
?
AS A RESULT, ONLY BIG-BEAKED INDIVIDUALS GOT ADEQUATE FOOD.

THE FINCHES WITH SMALLER BEAKS EITHER STARVED OR WERE TOO MALNOURISHED TO REPRODUCE.
BY THE NEXT GENERATION NATURAL SELECTION HAD INCREASED THE SIZE OF THE FINCHES' BEAKS BY TEN PERCENT.
BODY MASS ALSO INCREASED. A PERFECT DEMONSTRATION OF EVOLUTIONARY CHANGE.
EVOLUTION OFFERS THE ONLY SCIENTIFICALLY TESTABLE EXPLANATION
OF HOW THE NATURAL WORLD CAN PRODUCE SUCH A MULTIPLICITY OF FLORA AND FAUNA OUT OF SIMPLE AND UNDERSTANDABLE PROCESSES
TO GENERATE THE TREE OF LIFE.
END

FRACKING

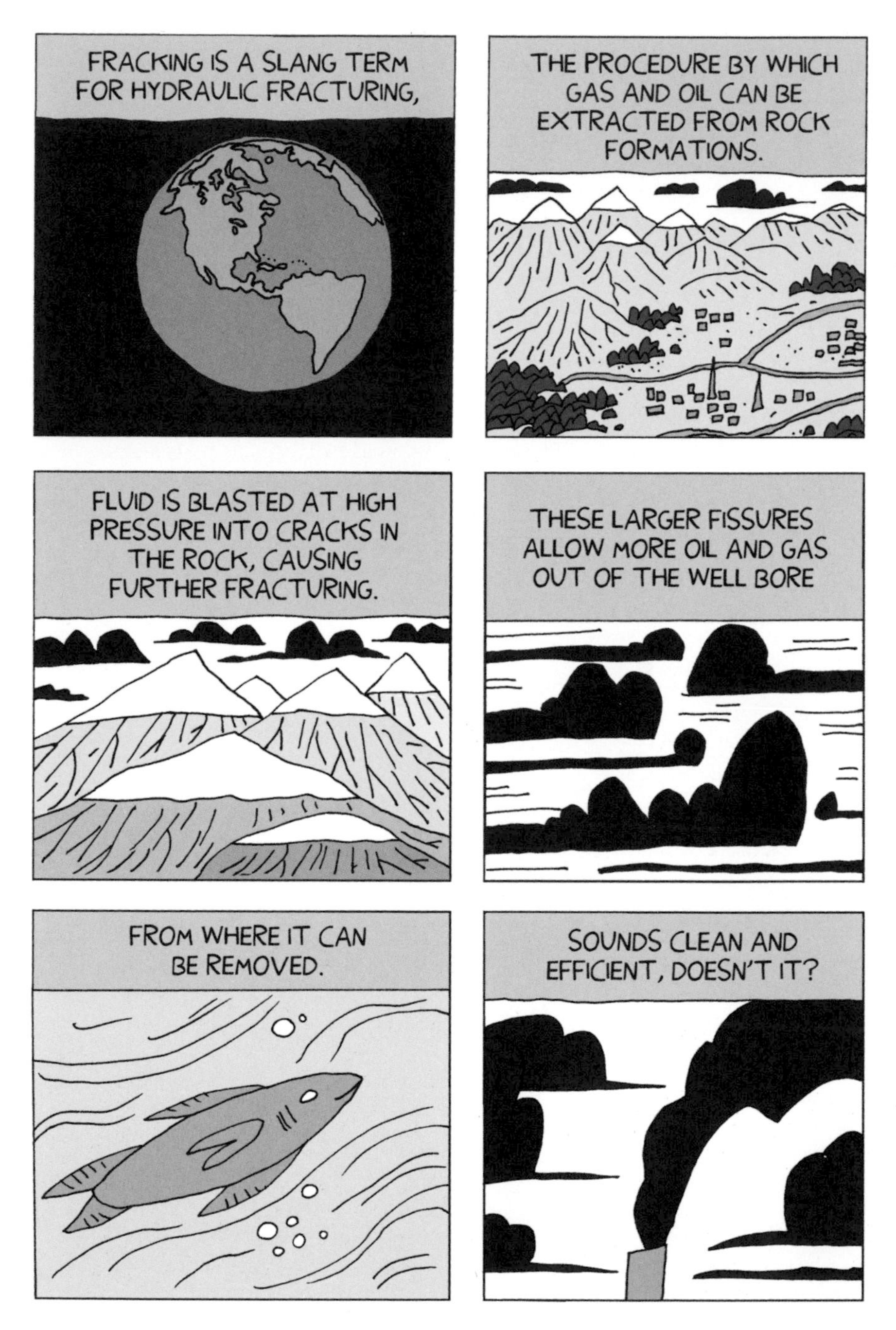
FRACKING IS A SLANG TERM FOR HYDRAULIC FRACTURING,
THE PROCEDURE BY WHICH GAS AND OIL CAN BE EXTRACTED FROM ROCK FORMATIONS.
FLUID IS BLASTED AT HIGH PRESSURE INTO CRACKS IN THE ROCK, CAUSING FURTHER FRACTURING.
THESE LARGER FISSURES ALLOW MORE OIL AND GAS OUT OF THE WELL BORE
FROM WHERE IT CAN BE REMOVED.
SOUNDS CLEAN AND EFFICIENT, DOESN'T IT?

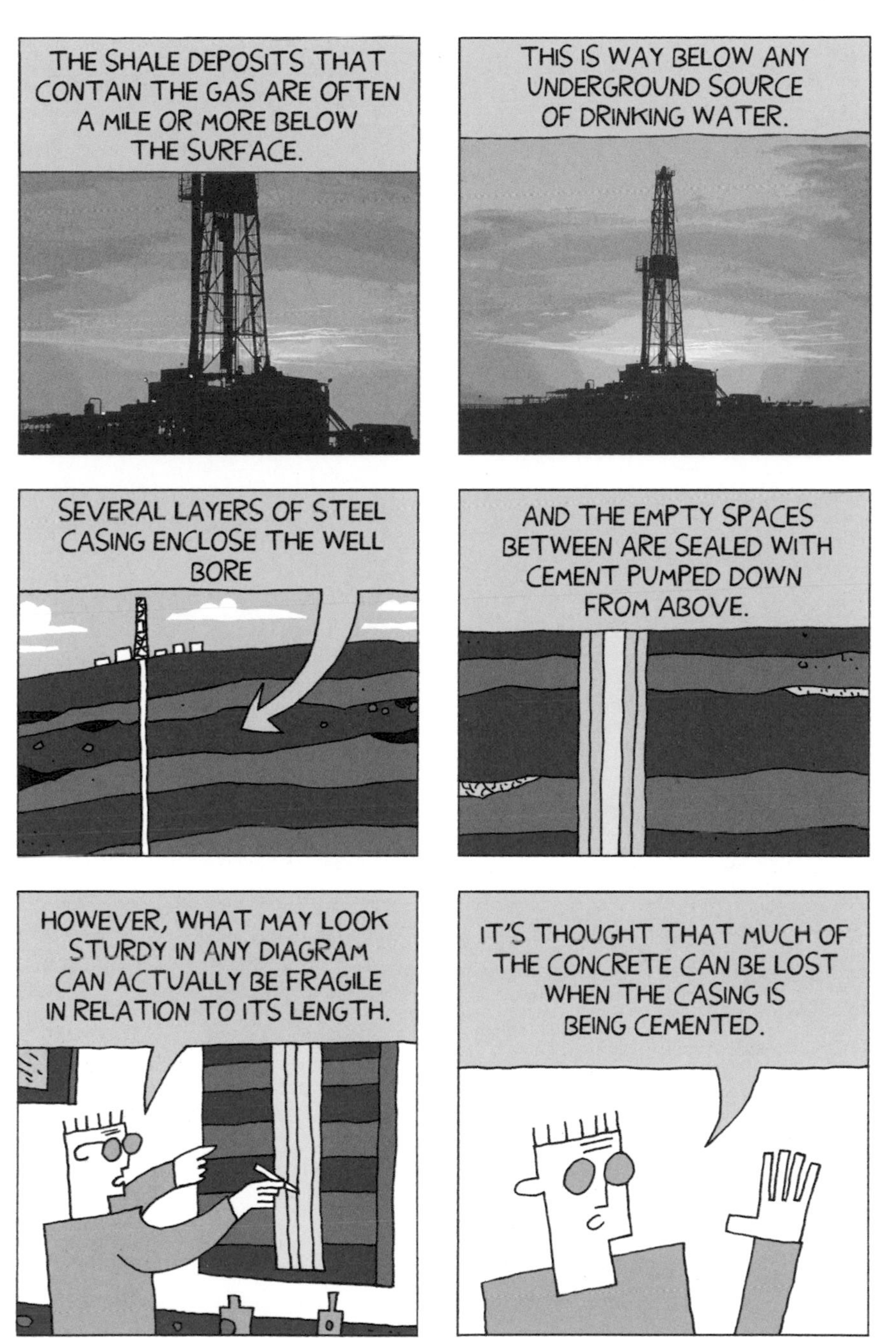
THE SHALE DEPOSITS THAT CONTAIN THE GAS ARE OFTEN A MILE OR MORE BELOW THE SURFACE.
THIS IS WAY BELOW ANY UNDERGROUND SOURCE OF DRINKING WATER.
SEVERAL LAYERS OF STEEL CASING ENCLOSE THE WELL BORE
AND THE EMPTY SPACES BETWEEN ARE SEALED WITH CEMENT PUMPED DOWN FROM ABOVE.
HOWEVER, WHAT MAY LOOK STURDY IN ANY DIAGRAM CAN ACTUALLY BE FRAGILE IN RELATION TO ITS LENGTH.
IT'S THOUGHT THAT MUCH OF THE CONCRETE CAN BE LOST WHEN THE CASING IS BEING CEMENTED.

AS A RESULT THERE MAY BE CONTAMINATION OF LOCAL GROUND WATER.

NOT JUST GAS MIGRATION, BUT ALSO TOXIC CHEMICALS.

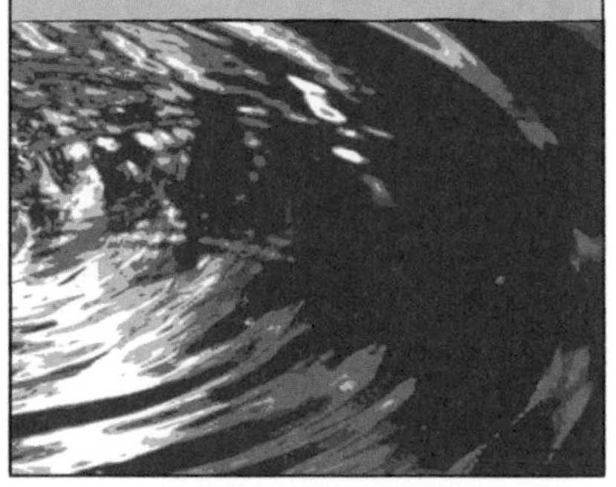

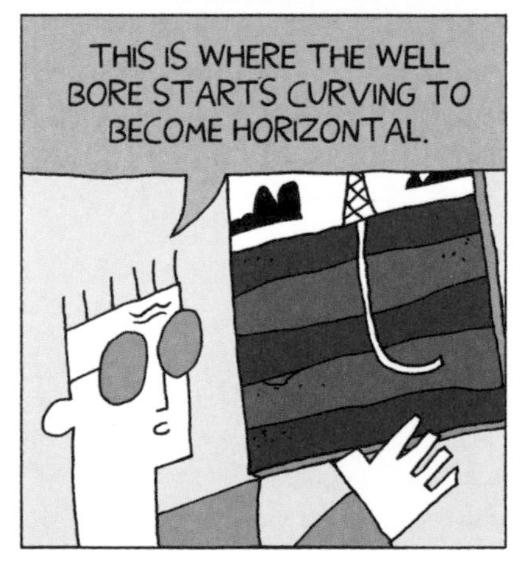

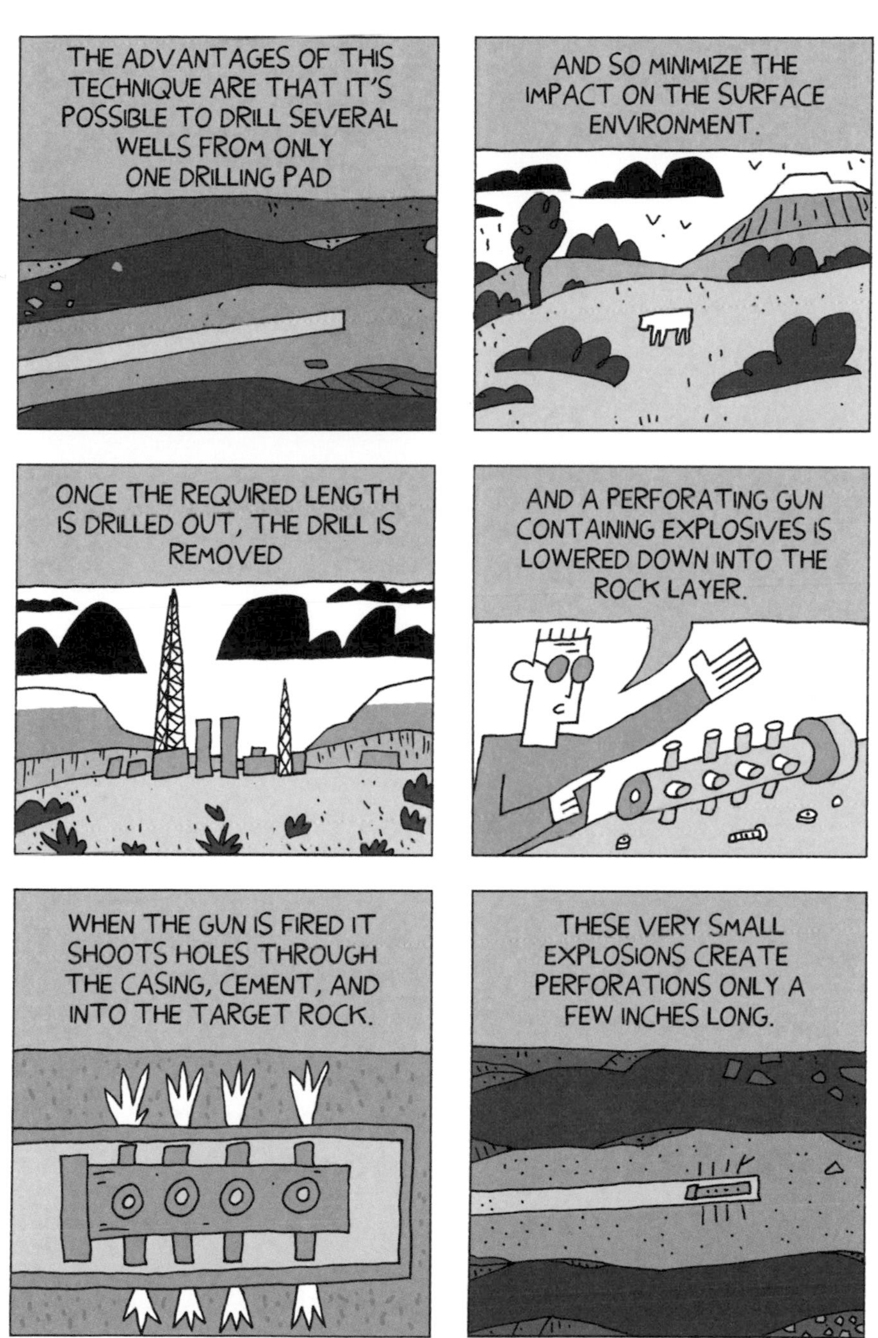
THE ADVANTAGES OF THIS TECHNIQUE ARE THAT IT'S POSSIBLE TO DRILL SEVERAL WELLS FROM ONLY ONE DRILLING PAD
AND SO MINIMIZE THE IMPACT ON THE SURFACE ENVIRONMENT.
ONCE THE REQUIRED LENGTH IS DRILLED OUT, THE DRILL IS REMOVED
AND A PERFORATING GUN CONTAINING EXPLOSIVES IS LOWERED DOWN INTO THE ROCK LAYER.
WHEN THE GUN IS FIRED IT SHOOTS HOLES THROUGH THE CASING, CEMENT, AND INTO THE TARGET ROCK.
THESE VERY SMALL EXPLOSIONS CREATE PERFORATIONS ONLY A FEW INCHES LONG.

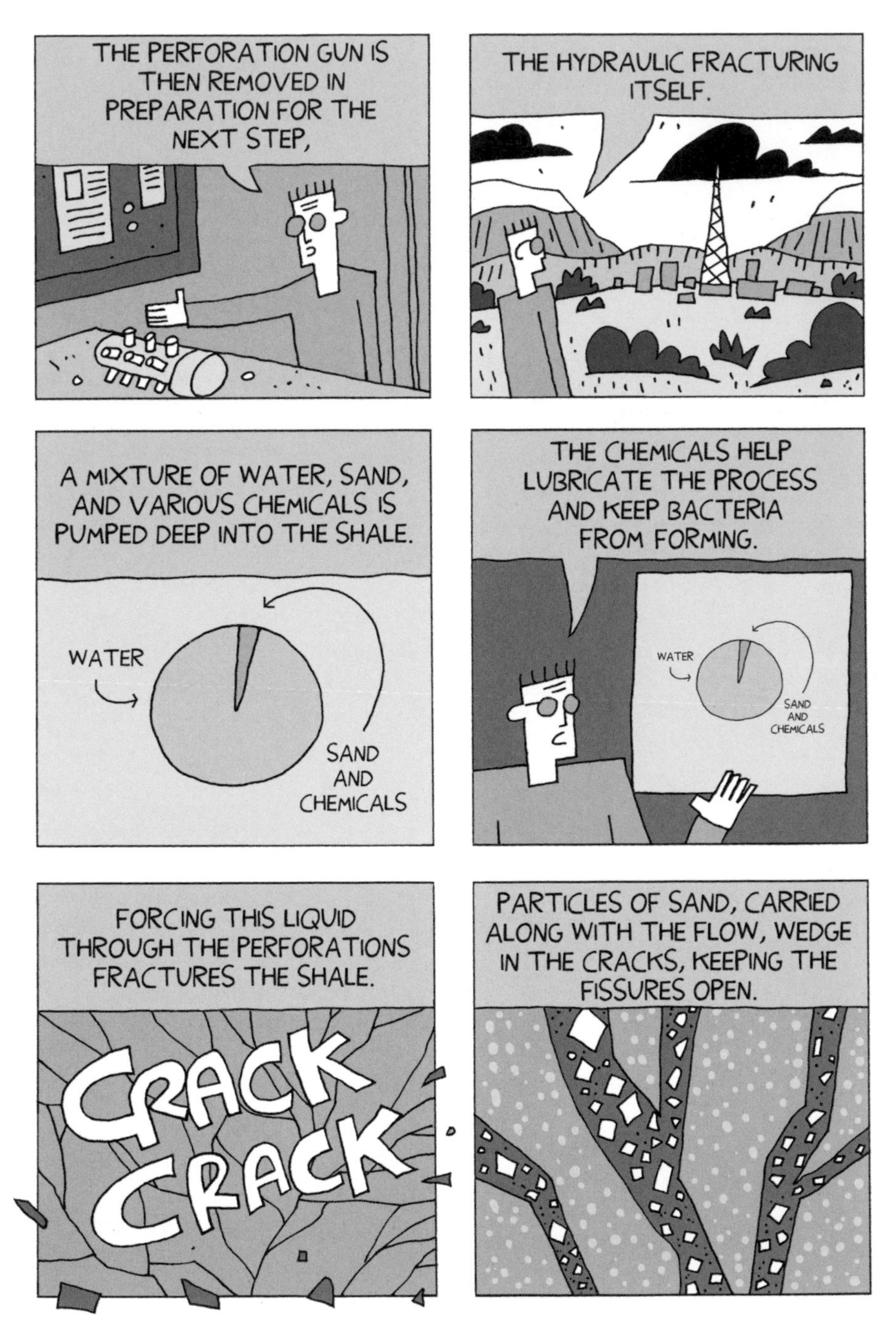
THE PERFORATION GUN IS THEN REMOVED IN PREPARATION FOR THE NEXT STEP,
THE HYDRAULIC FRACTURING ITSELF.
A MIXTURE OF WATER, SAND, AND VARIOUS CHEMICALS IS PUMPED DEEP INTO THE SHALE.
WATER
SAND AND CHEMICALS
THE CHEMICALS HELP LUBRICATE THE PROCESS AND KEEP BACTERIA FROM FORMING.
WATER
SAND AND CHEMICALS
FORCING THIS LIQUID THROUGH THE PERFORATIONS FRACTURES THE SHALE.
CRACK
CRACK
PARTICLES OF SAND, CARRIED ALONG WITH THE FLOW, WEDGE IN THE CRACKS, KEEPING THE FISSURES OPEN.

THE FRACTURING IS REPEATED ALONG THE ENTIRE HORIZONTAL LENGTH OF THE WELL, WHICH CAN EXTEND FOR SEVERAL MILES.

AT THIS POINT, AS MUCH OF THE FLUID AS POSSIBLE IS PUMPED OUT AND THE EXTRACTION OF THE GAS BEGINS.

IT'S AN INSPIRED TECHNIQUE. IN RESEARCHING THIS SUBJECT, I COULDN'T HELP BUT ADMIRE THE INGENUITY INVOLVED.

FRACKING HAS BROUGHT ABOUT THE LARGEST NATURAL GAS DRILLING BOOM IN THE HISTORY OF THE U.S.
IT HAS ALLOWED ACCESS TO VAST DEPOSITS OF NATURAL GAS, ENOUGH IT IS CLAIMED, TO SUPPLY THE COUNTRY FOR DECADES.
THE DRILLING BOOM STRETCHES ACROSS 31 STATES AND UP INTO CANADA.
IF IT COULD BE CERTAIN THAT THE FRACKING PROCESS WOULDN'T HARM THE LANDSCAPE, THEN ALL WOULD BE WELL.
HOWEVER, THERE ARE SERIOUS ENVIRONMENTAL CONCERNS TO CONSIDER.
THESE CONCERNS INCLUDE GROUNDWATER CONTAMINATION

THROUGH THE MIGRATION OF GASES AND FRACTURING CHEMICALS TO THE SURFACE,
THE MISHANDLING OF WASTE, AND THE RISKS TO AIR QUALITY.
THE CHEMICALS USED IN THE FRACKING PROCESS ARE THEMSELVES A SERIOUS DANGER.
THIS IS TRUE EVEN THOUGH THE CHEMICALS MAKE UP ONLY A SMALL AMOUNT OF THE LIQUID USED,
TYPICALLY ONLY 0.5 PERCENT OF THE TOTAL VOLUME OF THE FLUID.
THIS DOESN'T SOUND LIKE MUCH, I KNOW, BUT MILLIONS OF GALLONS OF WATER ARE INJECTED INTO THE GROUND DURING THIS PROCESS.

THIS MEANS THAT THE ACTUAL AMOUNT OF CHEMICAL PER ACTUAL OPERATION IS VERY LARGE.
FOR EXAMPLE, A FOUR MILLION GALLON FRACTURING OPERATION WOULD USE FROM 80 TO 330 TONS OF CHEMICALS.
HERE ARE SOME OF THE MOST COMMON CHEMICALS USED IN THE FRACKING PROCESS.
METHANOL: FOUND IN ANTI-FREEZE, PAINT SOLVENT, AND VEHICLE FUEL.
BENZENE: COMMONLY FOUND IN GASOLINE. LONG-TIME EXPOSURE CAN CAUSE CANCER, BONE MARROW FAILURE, OR LEUKEMIA.
YUM!
LEAD: PARTICULARLY HARMFUL TO CHILDREN'S NEUROLOGICAL DEVELOPMENT.
WANT SOME?
YAY!

HYDROGEN FLUORIDE: FOUND IN RUST REMOVERS, ALUMINUM BRIGHTENERS, AND HEAVY DUTY CLEANERS.
ITS FUMES ARE HIGHLY IRRITATING, CORROSIVE, AND POISONOUS.
NAPHTHALENE: A CARCINOGEN FOUND IN MOTH BALLS.
INHALATION CAN CAUSE RESPIRATION TRACT IRRITATION, NAUSEA, VOMITING, ABDOMINAL PAIN, FEVER, OR DEATH.
SULFURIC ACID: CORROSIVE TO ALL BODY TISSUES.
THE LETHAL DOSE IS BETWEEN ONE TEASPOONFUL AND ONE-HALF OUNCE.

FORMALDEHYDE: A CARCINOGEN FOUND IN EMBALMING AGENTS.

INGESTION OF EVEN AN OUNCE CAN CAUSE DEATH.

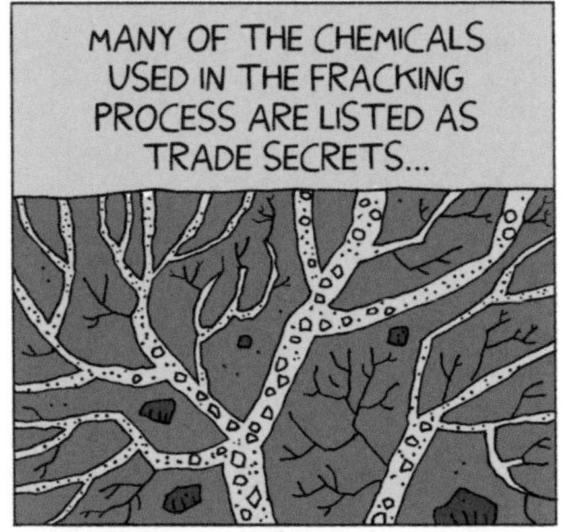
MANY OF THE CHEMICALS USED IN THE FRACKING PROCESS ARE LISTED AS TRADE SECRETS...

WHICH MEANS THAT THEIR EXACT NATURE IS UNKNOWN.

ONLY 25 TO 50 PERCENT OF THE TOXIC NON-BIODEGRADABLE MATERIAL IS RECOVERED.

THE REST IS SIMPLY LEFT THERE, INFUSED INTO THE LANDSCAPE, FOREVER.

WHAT HAPPENS TO THE LIQUID RECOVERED FROM THE DRILLING OPERATION?
READER'S VOICE
GOOD QUESTION. THE FIRST TYPE OF WASTEWATER IS CALLED BRINE.
THIS BRINE CONTAINS CUTTINGS FROM THE DRILLING PROCESS THAT CONTAIN MINERAL SALTS,
ARSENIC, MERCURY, THALLIUM, CHROMIUM, AND OTHER HEAVY METALS, ALONG WITH NATURALLY OCCURRING RADIOACTIVE MATERIALS.
BRINE IS GENERALLY DUMPED IN HUGE OPEN PITS LINED WITH PLASTIC.
THE FRACKING INDUSTRY SAYS THAT THEY ARE INCREASINGLY USING TANKS TO CONTAIN THIS BRINE.
DANGER
RESTRICTED
AREA

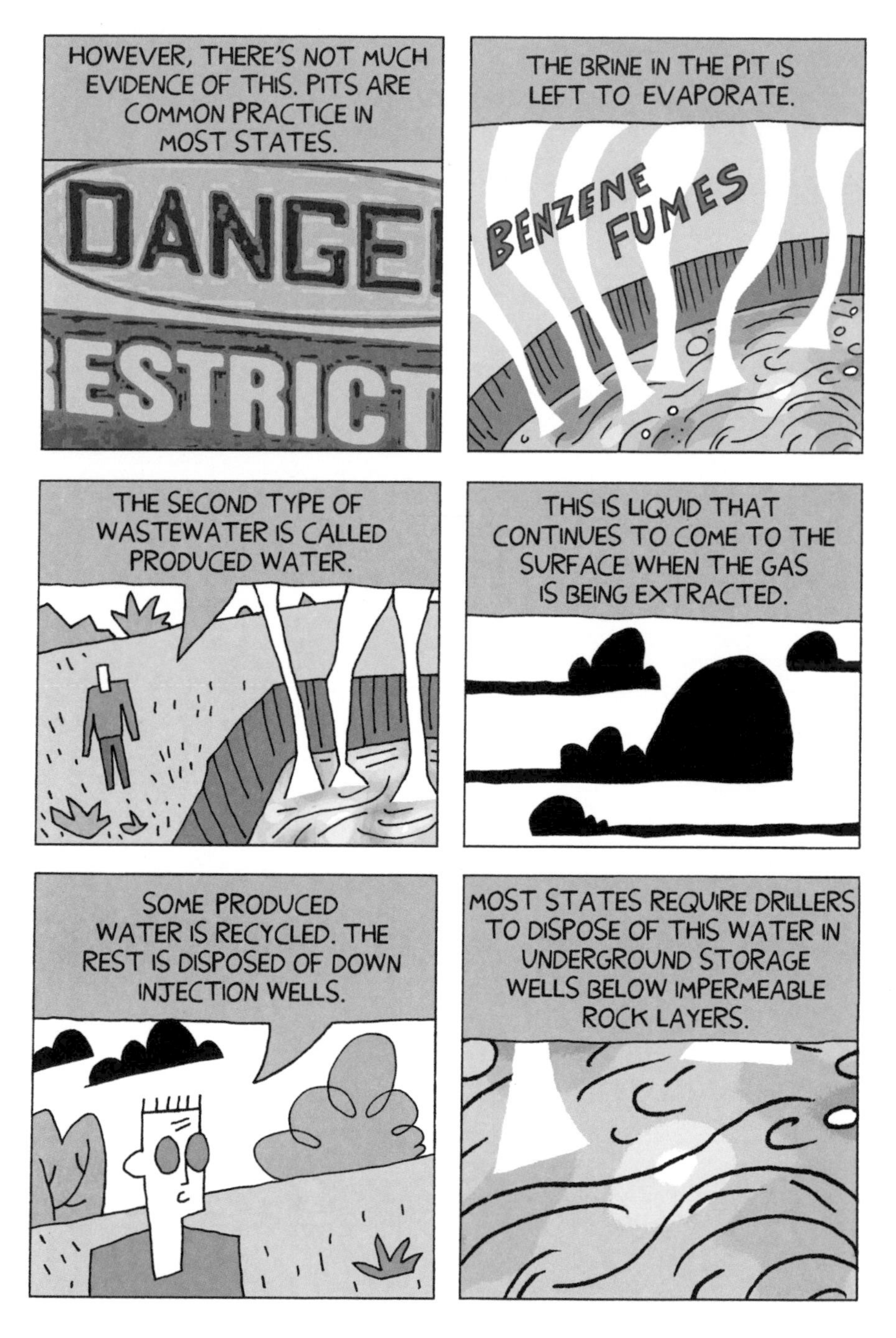
HOWEVER, THERE'S NOT MUCH EVIDENCE OF THIS. PITS ARE COMMON PRACTICE IN MOST STATES.
DANGE
ESTRICT
THE BRINE IN THE PIT IS LEFT TO EVAPORATE.
BENZENE FUMES
THE SECOND TYPE OF WASTEWATER IS CALLED PRODUCED WATER.
THIS IS LIQUID THAT CONTINUES TO COME TO THE SURFACE WHEN THE GAS IS BEING EXTRACTED.
SOME PRODUCED WATER IS RECYCLED. THE REST IS DISPOSED OF DOWN INJECTION WELLS.
MOST STATES REQUIRE DRILLERS TO DISPOSE OF THIS WATER IN UNDERGROUND STORAGE WELLS BELOW IMPERMEABLE ROCK LAYERS.

THE EXCEPTION IS PENNSYLVANIA.

IT'S THE ONLY STATE THAT ALLOWS DRILLERS TO DISCHARGE THEIR WASTE THROUGH SEWAGE TREATMENT PLANTS INTO RIVERS.

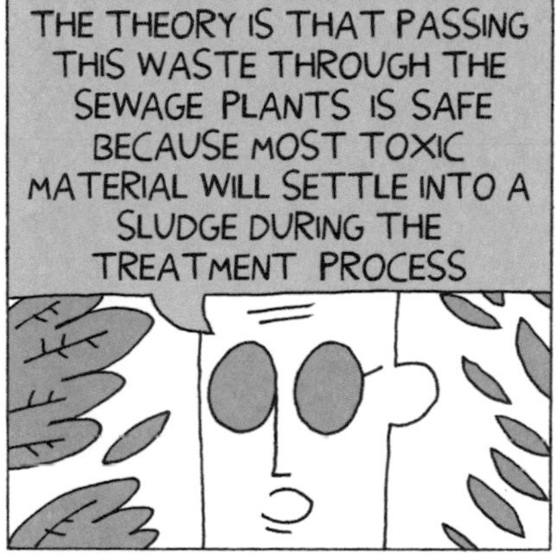
THE THEORY IS THAT PASSING THIS WASTE THROUGH THE SEWAGE PLANTS IS SAFE BECAUSE MOST TOXIC MATERIAL WILL SETTLE INTO A SLUDGE DURING THE TREATMENT PROCESS

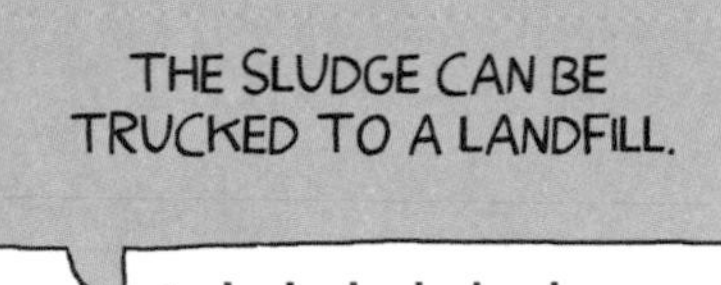
THE SLUDGE CAN BE TRUCKED TO A LANDFILL.

AND WHATEVER TOXIC MATERIAL REMAINS WILL BE DILUTED WHEN MIXED INTO RIVERS.

BUT SOME SEWAGE PLANTS WERE TAKING SUCH LARGE AMOUNTS OF WASTE WITH HIGH SALT LEVELS, THAT IN 2008

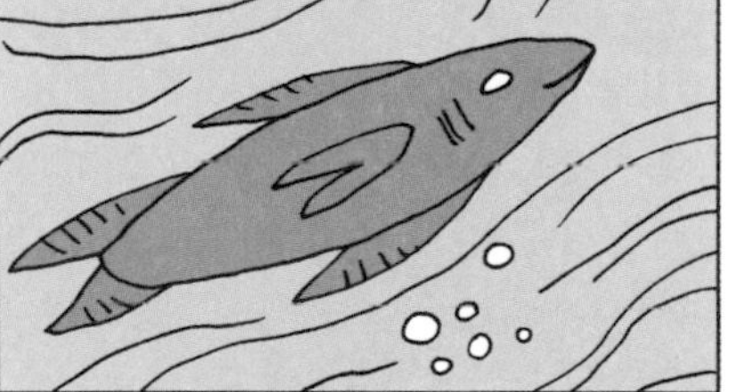

DOWNSTREAM UTILITIES BEGAN COMPLAINING THAT THE RIVER WATER WAS EATING AWAY AT THEIR MACHINES.
IT'S BROKEN.
ANOTHER PROBLEM WITH FRACKING IS CONTAMINATION BY NATURALLY OCCURRING RADIOACTIVE MATERIALS.
MOST SEWAGE TREATMENT PLANTS ARE NOT EQUIPPED TO REMOVE THIS MATERIAL
NOR ARE THEY REQUIRED BY LAW TO TEST FOR IT.
THESE LOW LEVELS OF RADIOACTIVITY POSE NO THREAT TO THE PUBLIC.
INDUSTRY SPOKESPERSON
IT'S MORE OF A PERCEPTION ISSUE THAN A REAL HEALTH THREAT.

IN 2011 THE NEW YORK TIMES REPORTED THAT OF THE MORE THAN 179 WELLS PRODUCING WASTEWATER
WITH HIGH LEVELS OF RADIATION
116 REPORTED LEVELS OF RADIOACTIVE MATERIAL 100 TIMES
THE LEVELS SET BY FEDERAL DRINKING WATER STANDARDS.
AT LEAST 15 OF THESE WELLS PRODUCED WASTEWATER MORE THAN 1,000 TIMES
THE AMOUNT OF RADIATION CONSIDERED ACCEPTABLE.

WE'RE INCREASINGLY RECYCLING WASTEWATER RATHER THAN DISPOSING OF IT.
EVEN WITH RECYCLING, THE AMOUNT OF WASTEWATER PRODUCED IN PENNSYLVANIA IS EXPECTED TO INCREASE.
ACCORDING TO INDUSTRY PROJECTIONS, 50,000 NEW WELLS ARE LIKELY TO BE DRILLED IN THE STATE OVER THE NEXT TWO DECADES.
IF YOU SCALE UP ALL THE GAS DRILLING ACTIVITY ACROSS THE ENTIRE UNITED STATES
AND LOOK AT THE IMMENSITY OF THE SHALE DEPOSITS UNDER THE LAND
SHALE DEPOSITS
THEN YOU HAVE TO ASK, WHERE CAN THE INEVITABLE VAST AMOUNTS OF WASTE SAFELY GO?

THE GAS INDUSTRY CLAIMS THAT THERE IS NO POSSIBILITY OF LEAKAGE AND CONTAMINATION
BECAUSE ANY WASTE MATERIAL IS DEEP BELOW IMPERMEABLE ROCK LAYERS.
THERE IS NO WAY WAY IT CAN RETURN TO THE SURFACE. THERE HAS NEVER BEEN AN EXAMPLE OF THIS HAPPENING.
WELL, I'VE SEEN FOOTAGE OF PEOPLE SETTING FIRE TO THEIR DRINKING WATER
BECAUSE THE WATER IS SO FULL OF FLAMMABLE METHANE GAS.
CHECK THIS OUT!
WOOAH!

THE TYPE OF METHANE OFTEN REPORTED IN DRINKING WATER IS BIOGENIC IN ORIGIN.

MEANING THAT IT WAS FORMED FROM DECAYING ORGANIC MATTER NEAR THE SURFACE AND HAD NOTHING TO DO WITH GAS DRILLING.

IT'S SIMPLY IMPOSSIBLE FOR THERMOGENIC GAS, THE KIND OF GAS WE DRILL FOR,

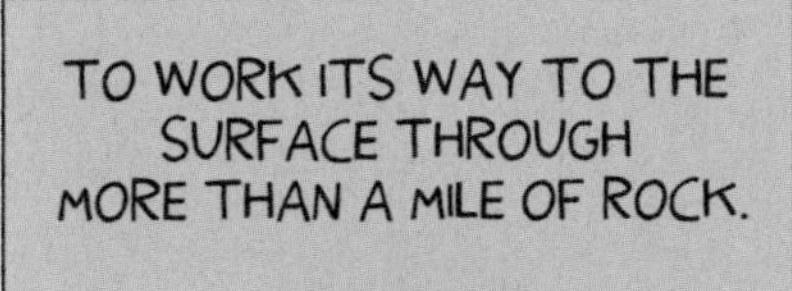
TO WORK ITS WAY TO THE SURFACE THROUGH MORE THAN A MILE OF ROCK.

YOU DON'T KNOW THAT BECAUSE NOT NEARLY ENOUGH RESEARCH HAS BEEN DONE TO CONFIRM OR DENY IT.

WE DON'T KNOW WHAT THE EFFECT OF SO MANY CLOSELY SPACED, RELATIVELY SHALLOW FRACKS COULD DO.

THE PROCESS COULD CREATE ROCK DISTURBANCES
THAT COULD OPEN UP PREVIOUSLY BLOCKED MIGRATION PATHS THROUGH JOINT SETS OR FAULTS.
ALTHOUGH THE ENVIRONMENTAL PROTECTION AGENCY HAS SINCE BACKED AWAY FROM ITS CLAIMS THAT IT HAD FOUND EVIDENCE OF CONTAMINATED GROUNDWATER IN PENNSYLVANIA...
AN INDEPENDENT STUDY BY ENVIRONMENTAL SCIENTISTS AT DUKE UNIVERSITY, DURHAM, NORTH CAROLINA, FOUND CLEAR EVIDENCE OF CONTAMINATION.
GLOOP!
BUBBLE
EVEN IF THE DEEP STORAGE OF THIS MATERIAL DID TURN OUT TO BE SAFE, THERE IS STILL THE ISSUE OF WELL CASING FAILURE TO CONSIDER.
A DRILLING WELL HAS TO BE DEEP, SOMETIMES AS MUCH AS A MILE DEEP AND TWO MILES OUT.

AS WE SAW, THE PIPE IS SHIELDED WITH CEMENT ALONG ITS ENTIRE LENGTH, PUMPED DOWN FROM ABOVE.
THE CEMENT IS SUPPOSED TO SEAL OFF GAS AND CONTAMINANTS FROM LEAKING INTO LOCAL AQUIFERS.
BUT CEMENT FAILURE IN THE OIL AND GAS INDUSTRY DOES HAPPEN.
CEMENT FAILURE CAUSED THE BLOWOUT IN THE GULF OF MEXICO.
THERE DOESN'T NEED TO BE LEAKAGE FROM DEEP UNDERGROUND FOR THERE TO BE CONTAMINATION.
IT CAN LEAK RIGHT OUT OF THE PIPE NEAR THE SURFACE.

ACCIDENTS DO HAPPEN, AND FRACKING REGULATION IS LAX.

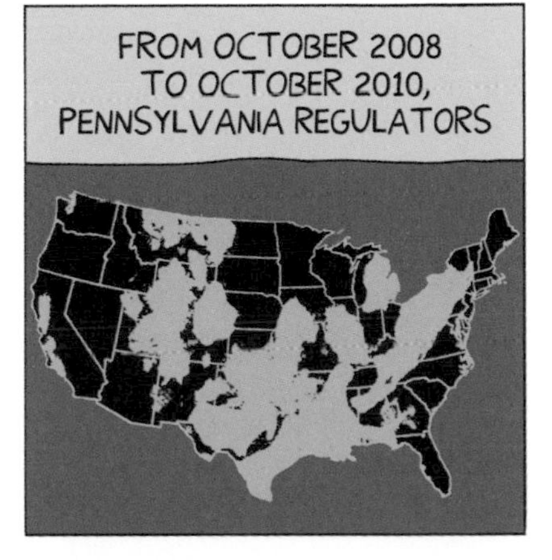
FROM OCTOBER 2008 TO OCTOBER 2010, PENNSYLVANIA REGULATORS

WERE MORE THAN TWICE AS LIKELY TO ISSUE A WRITTEN WARNING THAN TO LEVY A FINE

FOR ENVIRONMENTAL AND SAFETY VIOLATIONS, ACCORDING TO STATE DATA.

15 COMPANIES WERE FINED FOR DRILLING-RELATED VIOLATIONS IN 2008 AND 2009.

THESE COMPANIES PAID AN AVERAGE OF ABOUT $44,000 EACH IN FINES DURING THIS PERIOD...

LESS THAN HALF OF WHAT SOME COMPANIES EARNED IN PROFITS IN A DAY
AND A TINY FRACTION OF THE MORE THAN $2 MILLION SOME OF THEM PAID ANNUALLY TO HAUL AND TREAT THE WASTE.
TOM CORBET BECAME GOVERNOR OF PENNSYLVANIA IN 2011.
THIS MAN TOOK MORE CAMPAIGN CONTRIBUTIONS FROM THE GAS INDUSTRY THAN ALL HIS COMPETITORS COMBINED.
$
AS GOVERNOR, CORBET SAID HE WOULD REOPEN STATE LAND TO NEW DRILLING, REVERSING HIS PREDECESSOR'S DECISION.
HE WAS AGAINST A PROPOSED GAS EXTRACTION TAX ON THE INDUSTRY.
REGULATION OF THE INDUSTRY HAS BEEN TOO AGGRESSIVE.

I WILL DIRECT THE DEPARTMENT OF ENVIRONMENTAL PROTECTION TO SERVE AS A PARTNER WITH BUSINESS, COMMUNITIES, AND LOCAL GOVERNMENT.
IT SHOULD RETURN TO ITS CORE MISSION OF PROTECTING THE ENVIRONMENT BASED ON SOUND SCIENCE.
WHAT WE SEE IN PENNSYLVANIA IS A MICROCOSM OF WHAT IS HAPPENING ACROSS THE UNITED STATES.
FOR DECADES ENVIRONMENTAL PROTECTIONS HAVE BEEN ROLLED BACK
AT THE BEHEST OF THE OIL AND GAS INDUSTRY,
AN INDUSTRY NOW FREE TO GIVE VAST AMOUNTS OF CAMPAIGN CONTRIBUTIONS TO ITS CHOSEN POLITICIANS

WHOSE LOYALTY IS TO THEIR WEALTHY DONORS FIRST, RATHER THAN THE ELECTORATE.
THIS PROVISION WAS INSERTED AT THE BIDDING OF THEN VICE PRESIDENT DICK CHENEY.

AMONG THE MANY PROVISIONS OF THE 2005 FEDERAL ENERGY BILL WAS ONE THAT'S BECOME KNOWN AS THE HALLIBURTON LOOPHOLE.

CHENEY WAS A FORMER CHIEF EXECUTIVE OF HALLIBURTON: THE WORLD'S LARGEST PROVIDER OF FRACKING SERVICES.

THE HALLIBURTON LOOPHOLE STRIPPED THE ENVIRONMENTAL PROTECTION AGENCY OF ITS AUTHORITY TO REGULATE FRACKING.

IT ALSO CHANGED THE DEFINITION OF THE WORD POLLUTANT, SO THAT "MATERIAL INJECTED INTO A WELL TO FACILITATE THE PRODUCTION OF OIL OR GAS" COULD NO LONGER BE CONSIDERED A POLLUTANT.

THAT THERE SHOULD BE SUCH NAKED INTERFERENCE IN THE POLITICAL PROCESS BY BIG BUSINESS IS CLEARLY WRONG.
IF REGULATORS HAD EASY ACCESS TO INFORMATION AND SUFFICIENT MANPOWER TO ASSESS RISKS,
COMPARED TO ITS ENVIRONMENTAL COSTS.

IT IS SCIENCE, UNAFFECTED BY POLITICAL BIAS OR EMOTION, THAT SHOULD DECIDE WHETHER FRACKING IS ENVIRONMENTALLY SAFE OR NOT.

THEN WE, THE GENERAL PUBLIC, WOULD HAVE A CLEARER UNDERSTANDING OF THE BENEFITS OF FRACKING

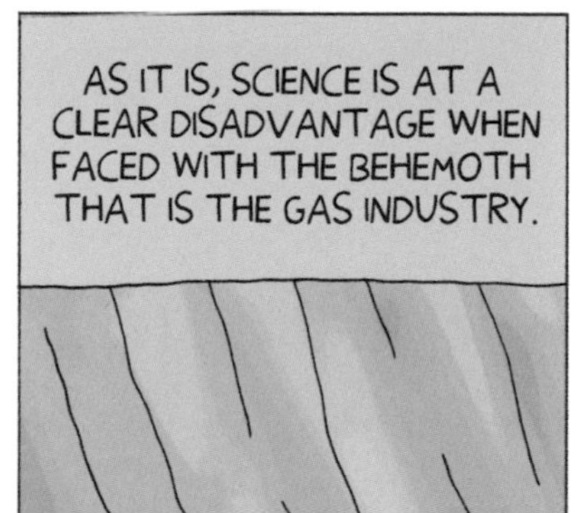
AS IT IS, SCIENCE IS AT A CLEAR DISADVANTAGE WHEN FACED WITH THE BEHEMOTH THAT IS THE GAS INDUSTRY.

THE INDUSTRY, WITH ALL ITS POLITICAL AND LEGAL MUSCLE AND ITS VAST FINANCIAL RESOURCES
SO THAT ANY CRITICISM OF FRACKING, HOWEVER SLIGHT, LOOKS LIKE LIBERAL BIAS.
THE TRUTH WILL EVENTUALLY SURFACE.

HAS POISONED DEBATE ON THE SUBJECT

HOWEVER, THE INDUSTRY IS NOT ALL-POWERFUL, BECAUSE AS THE TOBACCO COMPANIES FOUND OUT BEFORE IT.

KAFF! KAFF!

NO ONE CAN BURY THE TRUTH FOREVER.
END

CLIMATE CHANGE

IN ENGLAND, WHERE I LIVE, THE WINTER OF 2009-2010 WAS UNUSUALLY COLD.
IN FACT, THE NORTHERN HALF OF EUROPE EXPERIENCED ITS COLDEST WINTER SINCE 1981-1982.
THIS, FOR MANY, SHOWED THAT GLOBAL WARMING WAS NONSENSE.
BY CONTRAST, IN RUSSIA, THE SUMMER OF 2010 BROUGHT WITH IT A RECORD-BREAKING HEATWAVE.
FOREST FIRES AND DROUGHT RAVAGED THE COUNTRY.

THE NORTHERN EUROPE WINTER HAS BEEN CITED AS EVIDENCE THAT CLIMATE CHANGE IS A HOAX.
WHILE THE RUSSIAN SUMMER HAS BEEN CITED AS EVIDENCE THAT CLIMATE CHANGE IS REAL.
THE TRUTH IS THAT NEITHER OF THESE EVENTS CAN BE USED TO PROVE THE CASE EITHER WAY.
HUGE AS THESE EXTREME WEATHER EVENTS MAY APPEAR TO US ON OUR TINY HUMAN SCALE,
THEY ARE STILL MERE LOCAL EVENTS. IN ORDER TO GET ANY REAL UNDERSTANDING OF THE PLANET'S CLIMATE,
YOU HAVE TO LOOK AT WEATHER SYSTEMS GLOBALLY OVER A LONG PERIOD OF TIME.

ICE CORE SAMPLES FROM GREENLAND, ANTARCTICA, AND MOUNTAIN GLACIERS
Queen Elizabeth Is (Canada)
Arctic Ocean
Barents Sea
Nord
Svalbard Is (Norway)
Greenland Sea
NORWAY
Baffin Bay
Daneborg
Upernavik
Norwegian Sea
Scoresbysund
Godhavn
Davis Strait
Denmark Strait
ICELAND
Hudson Strait
Mt Gunnbjørn
NUUK
GREAT
HAVE GIVEN SCIENTISTS INFORMATION ON THE EARTH'S CLIMATE GOING BACK THOUSANDS OF YEARS.
IN THE LAST 650, 000 YEARS THERE HAVE BEEN SEVEN CYCLES OF GLACIAL ADVANCE AND RETREAT.
MOST OF THESE CHANGES WERE DUE TO SMALL SHIFTS IN THE EARTH'S ORBIT.
INFORMATION ON THE PRESENT STATE OF THE CLIMATE COMES FROM EARTH-ORBITING SATELLITES AND OTHER TECHNOLOGICAL ADVANCES.
FROM THIS INFORMATION THE EVIDENCE FOR RAPID CLIMATE CHANGE IN MODERN TIMES IS COMPELLING.

GLOBAL SEA LEVELS HAVE RISEN ABOUT 7 INCHES (17 CM) IN THE PAST CENTURY.
A RATE OF INCREASE THAT HAS DOUBLED IN THE PAST DECADE.
THERE HAS BEEN A CONSISTENT GLOBAL SURFACE TEMPERATURE RISE SINCE THE 1880s
AND MOST OF THIS WARMING HAS OCCURRED SINCE THE 1970s
WITH TWO OF THE WARMEST YEARS HAPPENING IN THE PAST 12 YEARS.
ALL THIS HAS TAKEN PLACE EVEN THOUGH THE 2000s HAVE EXPERIENCED A SOLAR OUTPUT DECLINE.

THE OCEANS HAVE ABSORBED MUCH OF THE INCREASED HEAT, WITH THE TOP 2,300 FEET (ABOUT 706 METERS)
SHOWING WARMING OF 0.302 DEGREES FAHRENHEIT SINCE 1969.
BOTH THE GREENLAND AND ANTARCTIC ICE SHEETS HAVE DECREASED IN MASS.
YES, THAT'S RIGHT! DATA FROM NASA'S GRAVITY RECOVERY AND CLIMATE EXPERIMENT SHOWS THAT
GREENLAND LOST 36 TO 60 CUBIC MILES (150 TO 250 CUBIC KILOMETERS) OF ICE PER YEAR BETWEEN 2002 AND 2006.
WHILE ANTARCTICA LOST ABOUT 36 CUBIC MILES (152 CUBIC KILOMETERS) OF ICE BETWEEN 2002 AND 2005.

IN JULY 2012 AN ICEBERG TWICE THE SIZE OF MANHATTAN BROKE AWAY FROM THE PETERMAN GLACIER IN GREENLAND.
WE SEE THESE CHANGES IN THE HIMALAYAS, THE ANDES, THE ROCKIES, ALASKA, AND AFRICA.
THAT THERE'S BEEN SUCH AN OBVIOUS JUMP IN GLOBAL TEMPERATURE DURING INDUSTRIAL TIMES
DOES SUGGEST THAT THESE CHANGES ARE HUMAN-MADE.
BUT WHAT IS THE SCIENCE BEHIND THIS THEORY?
THE ARGUMENT FOR HUMAN-DRIVEN CLIMATE CHANGE IS AS FOLLOWS.

THE EARTH'S ATMOSPHERE KEEPS THE PLANET WARMER THAN IT WOULD BE IF IT DIDN'T HAVE AN ATMOSPHERE.
IONOSPHERE
MESOSPHERE
OZONE
STRATOSPHERE
TROPOSPHERE
THE MAIN GASES THAT CONTRIBUTE TO THIS EFFECT ARE CARBON DIOXIDE, METHANE, AND WATER VAPOR.
THEIR ABILITY TO ACT AS GREENHOUSE GASES CAN BE DEMONSTRATED IN A LABORATORY.
SINCE THE INDUSTRIAL REVOLUTION, THE QUANTITY OF THESE GREENHOUSE GASES IN THE ATMOSPHERE HAS INCREASED SHARPLY.
HUMAN ACTIVITY HAS POURED THESE GASES INTO THE ATMOSPHERE.
THE TEMPERATURE OF THE EARTH'S ATMOSPHERE HAS BEEN RISING, AND IT CONTINUES TO RISE.

AND LASTLY, THE INCREASE IN GLOBAL TEMPERATURE CORRELATES WITH THE INCREASE OF GREENHOUSE GASES.
SO YOU SAY, BUT MANY ARE SKEPTICAL OF THIS ARGUMENT.
THAT'S TRUE. THERE ARE TWO KINDS OF PEOPLE WHO DISAGREE WITH THE THEORY OF HUMAN-MADE CLIMATE CHANGE.
THERE ARE THOSE WHO DON'T KNOW ALL THE INFORMATION AND ARE THEREFORE DOUBTFUL OF THE THEORY.
BUT THEY TEND TO BE AWARE OF THE LIMITS OF THEIR KNOWLEDGE
AND SO REMAIN OPEN TO THE IDEA THAT CLIMATE CHANGE MIGHT BE REAL.

THEN THERE ARE THOSE WHO SIMPLY REJECT CLIMATE CHANGE, NOT ON SCIENTIFIC GROUNDS, BUT ON THE GROUNDS OF IDEOLOGY AND DOGMA.
"THE GREATEST HOAX EVER PERPETRATED ON THE AMERICAN PEOPLE."
JAMES INHOFE
U.S. SENATOR, OKLAHOMA
MOST OF THE CLIMATE CHANGE DENIERS ON THE INTERNET TEND TO POPULATE MILITANTLY RIGHT-WING BLOGS, INHABITED ENTIRELY BY OTHER DENIERS
WHERE A SMATTERING OF CITATIONS FROM LEGITIMATE SCIENTIFIC AUTHORITIES
ARE USED TO BOLSTER UP THIN ARGUMENTS AND OUTRIGHT DISTORTIONS.
CLIMATE CHANGE HOAX
BUT ISN'T IT TRUE THAT A GROWING NUMBER OF EMINENT SCIENTISTS NOW BELIEVE CLIMATE CHANGE TO BE WRONG?

IN AN ARTICLE PUBLISHED IN THE PROCEEDINGS OF THE NATIONAL ACADEMY OF SCIENCES (2010),
A GROUP OF SCHOLARS DID A STATISTICAL BREAKDOWN OF THE OPINIONS OF THE WORLD'S MOST PROMINENT CLIMATE EXPERTS.
PNAS
THEIR CONCLUSIONS WERE THAT THOSE SKEPTICAL OF THE EVIDENCE OF HUMAN-MADE CLIMATE CHANGE...
COMPRISED ONLY 2.5 PERCENT OF THE WORLD'S TOP 200 CLIMATE SCIENTISTS.
A TINY SLIVER OF FRINGE OPINION.
IT'S ONE THING TO BE SKEPTICAL, BUT IT'S ANOTHER THING ENTIRELY TO BELIEVE IN A CONSPIRACY.

"EUROPEANS NOW SEE GLOBAL WARMING AS A MEANS OF HAMPERING U.S. ECONOMIC COMPETITIVENESS."
STEVE MILLOY
FOX NEWS COLUMNIST
A CONSPIRACY THEORY CAN BE DEFINED AS ANY WORLD VIEW THAT TRACES IMPORTANT EVENTS
TO A SECRETIVE, NEFARIOUS CABAL, AND WHOSE PROPONENTS RESPOND TO CONTRARY FACTS, NOT BY MODIFYING THEIR HYPOTHESES
BUT INSTEAD BY INSISTING ON THE EXISTENCE OF EVER-WIDENING CIRCLES
OF HIGH-LEVEL CONSPIRATORS, WHO CONTROL MOST OR ALL PARTS OF SOCIETY.
RADICALIZED CLIMATE CHANGE DENIERS BELIEVE THAT GLOBAL WARMING IS A BACK-DOOR METHOD OF CREATING A SOCIALIST STATE.
GREENPEACE

BUT AREN'T YOU ALSO SUGGESTING A CONSPIRACY?

THERE IS REAL OPPOSITION TO THE SCIENTIFIC CONSENSUS, AND THIS OPPOSITION IS CONCENTRATED AROUND POLITICAL MOVEMENTS WITH STRONG TIES TO THE COAL AND OIL INDUSTRY.

EXXONMOBIL HAS POURED MILLIONS OF DOLLARS INTO CLIMATE CHANGE DENIAL GROUPS:

THE HEARTLAND INSTITUTE, THE HERITAGE FOUNDATION, THE GEORGE C. MARSHALL INSTITUTE

THE AMERICAN ENTERPRISE INSTITUTE. ALL GROUPS THAT ARE AT THE HEART OF CLIMATE CHANGE DENIAL. ANOTHER COMPANY THAT FUNDS SUCH GROUPS IS KOCH INDUSTRIES.

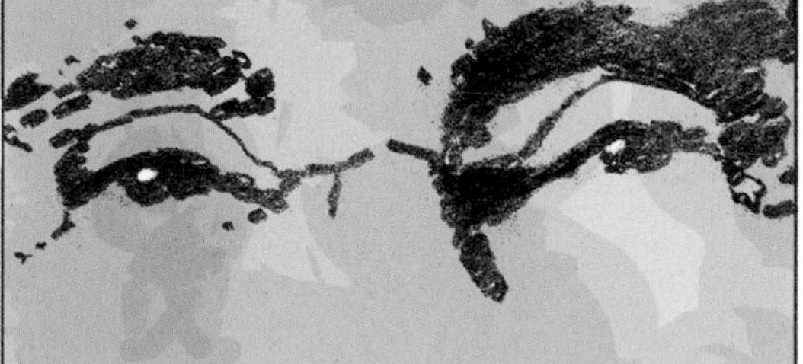

KOCH INDUSTRIES IS OWNED BY THE BILLIONAIRE KOCH BROTHERS, CHARLES AND DAVID KOCH.

THE COMPANY RUNS OIL REFINERIES, COAL PLANTS, AND LOGGING FIRMS AMONG OTHER BUSINESSES.
KOCH INDUSTRIES HAS ANNUAL REVENUES OF ROUGHLY 100 BILLION DOLLARS A YEAR. CHARLES AND DAVID ARE WORTH $25 BILLION EACH.
THE KOCHS WANT TO PAY LESS TAX, KEEP MORE PROFIT, AND BE RESTRAINED BY LESS REGULATION.
1 FOR YOU, 19 FOR ME
TAXED ENOUGH ALREADY
YES WE CAN'T
DON'T TAX ME BRO!
THEY ARE MOTIVATED BY AN IDEOLOGICAL COMMITMENT TO MINIMAL GOVERNMENT AND FREE MARKETS.
AMERICANS FOR PROSPERITY IS ONE OF SEVERAL GROUPS SET UP BY THE KOCH BROTHERS TO PROMOTE THEIR POLITICS.
GLOBAL WARMING
THE AFP HAVE TOURED THE U.S., ORGANIZING RALLIES AGAINST ATTEMPTS TO TACKLE CLIMATE CHANGE.
GLOBAL WARMING IS A SOCIALIST SCAM

THE AFP ALSO PROVIDED THE KEY ORGANIZING TOOLS THAT SET THE TEA PARTY RUNNING.
1 FOR YOU. 19 FOR ME
TAXED ENOUGH ALREADY
YES WE CAN'T
SOUNDS LIKE ANOTHER CRACK-POT CONSPIRACY THEORY TO ME.
IT'S SIMPLY THE REACTION YOU'D EXPECT FROM BIG BUSINESS WHEN ITS INTERESTS ARE THREATENED.
THE OIL AND COAL INDUSTRIES HAVE A LOT AT STAKE.
IF THE ANSWER TO THE PROBLEM OF CLIMATE CHANGE INVOLVES PHASING OUT FOSSIL FUELS, THEN A DIFFERENT SET OF PEOPLE WILL BE MAKING MONEY.
IT'S NO WONDER THAT THEY'RE FIGHTING TOOTH AND NAIL AGAINST IT.

BUT WHAT ABOUT THIS CLIMATEGATE BUSINESS? WASN'T CLIMATEGATE PROOF THAT ENVIRONMENTALISTS ALSO TWIST THE FACTS TO SUIT THEIR OWN AGENDA?
THE CLIMATEGATE EMAIL CONTROVERSY BEGAN IN NOVEMBER 2009
WHEN STOLEN EMAILS FROM THE UNIVERSITY OF EAST ANGLIA'S CLIMATE RESEARCH UNIT WERE MADE AVAILABLE TO READ ON THE INTERNET.
OUT OF THESE THOUSANDS OF EMAILS,
A HANDFUL OF STATEMENTS WERE SELECTED, THAT OUT OF CONTEXT, COULD BE SEEN AS CONTROVERSIAL,
AS IF SCIENTISTS AT THE UNIVERSITY HAD MANIPULATED RESEARCH IN ORDER TO SUPPORT PRECONCEIVED IDEAS ABOUT CLIMATE CHANGE.

SIX INDEPENDENT INVESTIGATIONS EXONERATED THE SCIENTISTS.
DESPITE SOME CONCERNS ABOUT DATA-SHARING AT THE CLIMATE RESEARCH CENTRE
SCIENTISTS HAD SHOWN INTEGRITY AND HONESTY.
Global Temperatures
Annual Average
Five Year Average
Temperature Anomaly (°C)
0.6
0.4
0.2
0
-0.2
-0.4
-0.6
1860 1880 1900 1920 1940 1960 1980 2000
NO SCIENCE HAD BEEN FALSIFIED, MANIPULATED, EXAGGERATED, OR FUDGED.
HOWEVER, THE DAMAGE HAD ALREADY BEEN DONE. IN A STUDY, SCOTT MANDIA, A METEOROLOGY PROFESSOR AT THE SUFFOLK COUNTY COMMUNITY COLLEGE, LONG ISLAND, USA, FOUND THAT MEDIA OUTLETS
HAD DEVOTED FIVE TO ELEVEN TIMES MORE STORIES TO THE ACCUSATIONS AGAINST THE SCIENTISTS THAN THEY HAD TO THE RESULTING EXONERATIONS.
CLIMATE "FRAUD"

A CLEAR CASE OF UNFAIR AND UNBALANCED REPORTING.
COMPARE THIS TO THE TINY AMOUNT OF MEDIA COVERAGE THAT KOCH INDUSTRIES RECEIVED
AFTER THEY WERE CAUGHT RED-HANDED GIVING MONEY TO GROUPS PROMOTING CLIMATE CHANGE AS A HOAX.
SO WHERE DOES THAT LEAVE US?
IT LEAVES US IN A PLACE WHERE BIG OIL CONTINUES TO DISTORT THE DEMOCRATIC PROCESS,
WHILE THE MEDIA LIES ASLEEP AT THE WHEEL.

MEANWHILE THE FUTURE LOOKS BLEAK.
DROUGHT, HUNGER, DISEASE, THE EXTINCTION OF AT LEAST A FOURTH OF THE WORLD'S SPECIES,
THE LOSS OF VITAL GLOBAL NATURAL AREAS LIKE THE AMAZON AND THE GREAT BARRIER REEF AND
THE FLOODING OF COASTS AND ISLANDS OCCUPIED BY HUNDREDS OF MILLIONS OF PEOPLE.
THE SCIENCE PREDICTS THAT THESE EVENTS COULD HAPPEN. LET'S NOT LEAVE IT TO THE SUPER-RICH TO DECIDE
WHO LIVES AND WHO DIES.
END

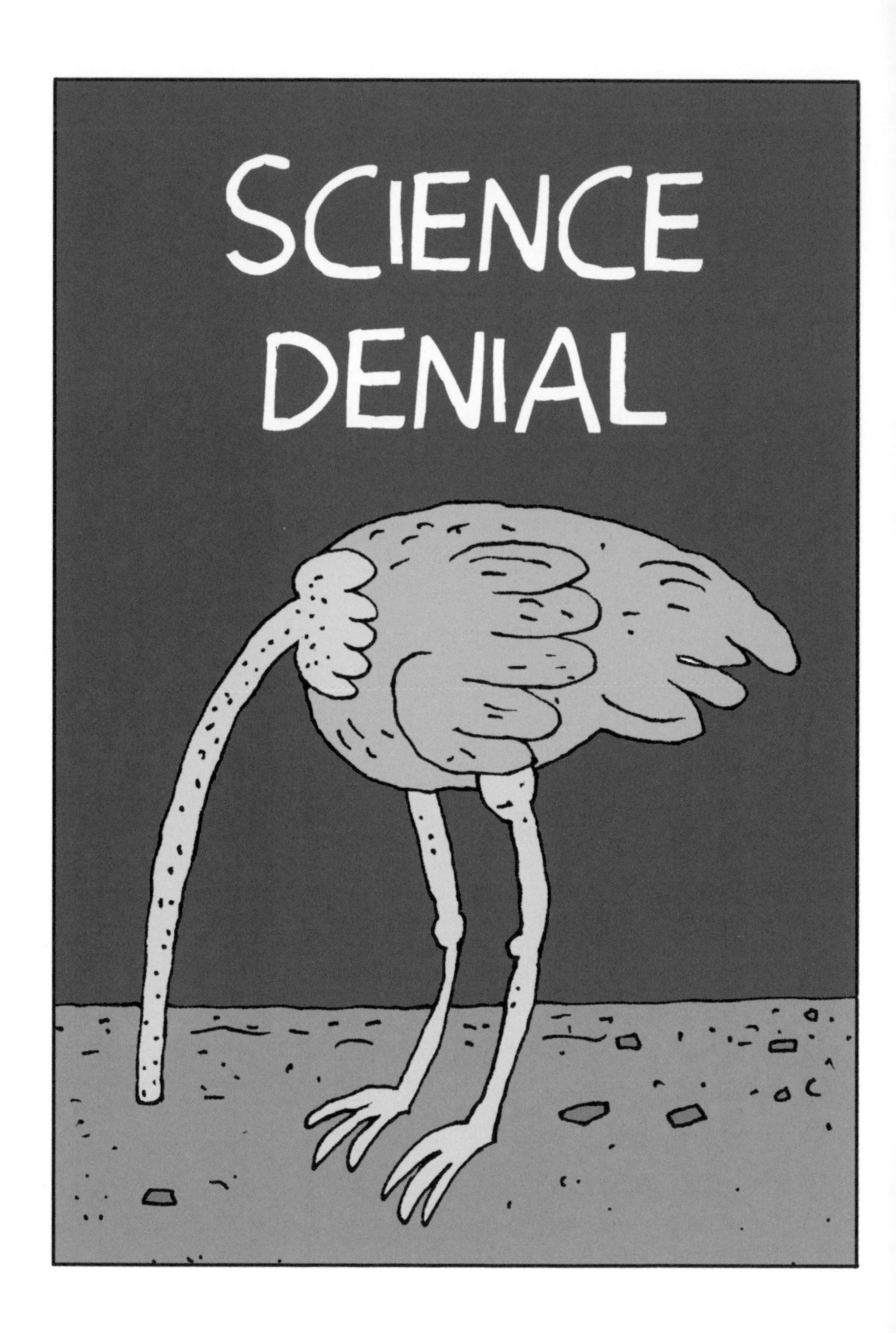
SCIENCE
DENIAL

THE AREAS OF SCIENCE THAT GENERATE THE FIERCEST DEBATE
TEND TO BE THOSE WE'RE OBLIGED TO TAKE ON TRUST.
THERE IS LITTLE ARGUMENT ANYMORE OVER THE SHAPE OF THE EARTH

OR THE ROLE OF MICRO-ORGANISMS IN DISEASE.
ACHOO!

BUT MORE DIFFICULT CONCEPTS, SUCH AS QUANTUM MECHANICS, NEED A HIGH LEVEL OF SPECIALIST KNOWLEDGE TO BE PROPERLY UNDERSTOOD,

SO THESE AREAS REMAIN THE DOMAIN OF SCIENTISTS.
ER!

HOWEVER, MANY OF SCIENCE'S CONCLUSIONS APPEAR TO CONTRADICT COMMON BELIEF
AND TO THREATEN IMPORTANT ASPECTS OF OUR LIVES.
HIV DOESN'T CAUSE AIDS. IT'S A BIG PHARMA CONSPIRACY.
THE WORLD WAS CREATED ONLY SIX THOUSAND YEARS AGO. EVERYONE KNOWS THAT.
CLIMATE CHANGE ISN'T HAPPENING, AND EVEN IF IT IS, IT'S NOT CAUSED BY HUMAN ACTIVITY.
SMOKING DOESN'T CAUSE CANCER. MY GRANDMOTHER LIVED UNTIL SHE WAS 103, AND SMOKED SINCE THE DAY SHE WAS BORN.

MY KID GOT AUTISM AFTER GETTING HIS SHOTS. THE VACCINE MUST HAVE DONE IT.

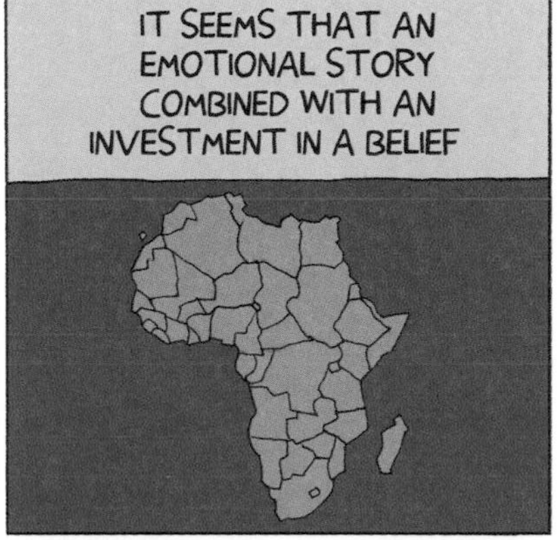
IT SEEMS THAT AN EMOTIONAL STORY COMBINED WITH AN INVESTMENT IN A BELIEF

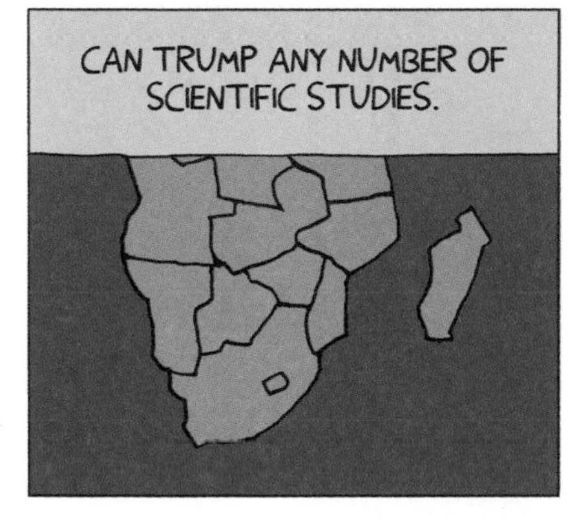
CAN TRUMP ANY NUMBER OF SCIENTIFIC STUDIES.

THIS CAN HAVE LETHAL CONSEQUENCES.

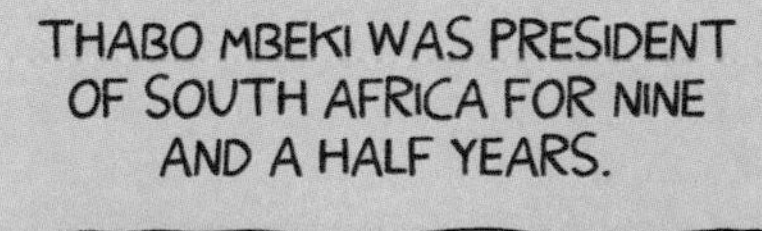
THABO MBEKI WAS PRESIDENT OF SOUTH AFRICA FOR NINE AND A HALF YEARS.

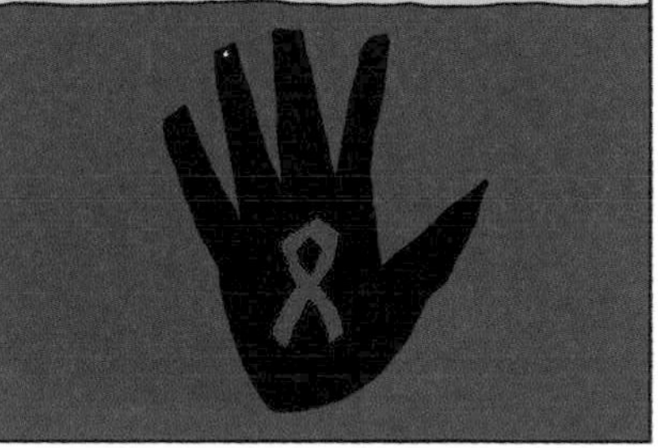
HIS DENIAL THAT HIV CAUSED AIDS PREVENTED THOUSANDS OF HIV-POSITIVE MOTHERS FROM RECEIVING ANTI-RETROVIRAL DRUGS

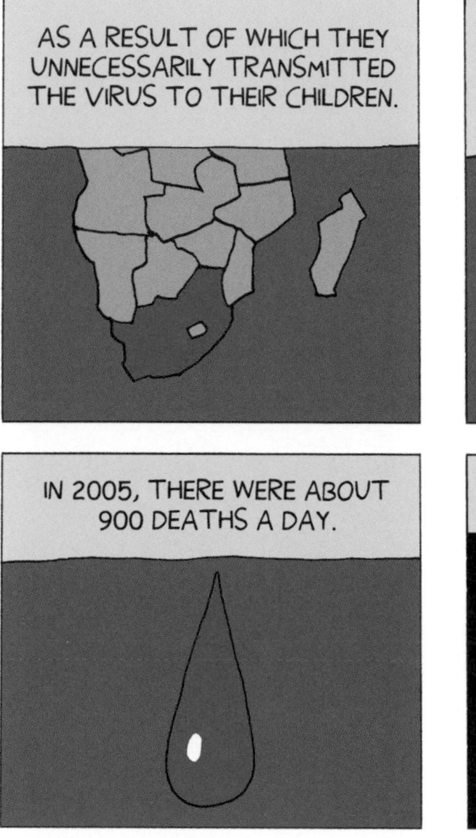

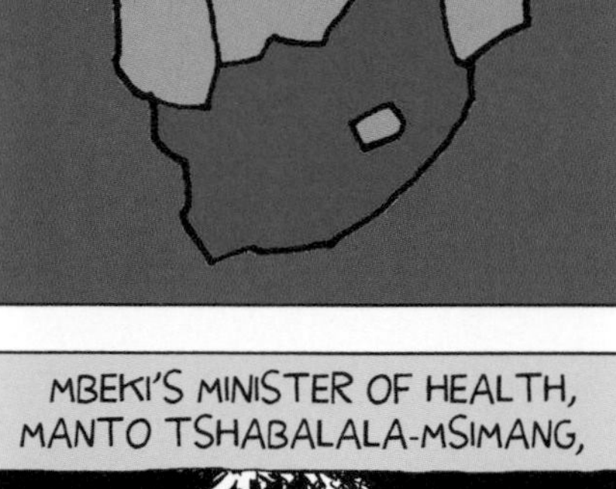

MBEKI'S MINISTER OF HEALTH, MANTO TSHABALALA-MSIMANG,

IN CONTRAST BOTH BOTSWANA AND NAMIBIA ACHIEVED GREAT RESULTS WITH THEIR PREVENTING MOTHER-TO-CHILD TRANSMISSIONS PROGRAMS.

THE HUMAN MIND IS NOTABLE FOR ITS ABILITY TO CLING TO ITS BELIEFS LONG PAST THE POINT
WHERE ANY EVIDENCE EXISTS TO SUPPORT THOSE BELIEFS.
A FEW SMOKES NEVER HURT ANYONE.
CORPORATE BUSINESS HAS BECOME EXPERT AT EXPLOITING THIS WEAKNESS IN HUMAN PSYCHOLOGY.
YOU HAVE CANCER.
THE STRATEGY OF CREATING DOUBT OVER THE VALIDITY OF SCIENTIFIC RESEARCH
WAS FIRST DEVELOPED BY THE TOBACCO INDUSTRY IN THE 1950s.
FOR OBVIOUS REASONS, IT WAS KEEN TO DISCREDIT THE SCIENTIFIC LINK BETWEEN SMOKING AND ITS HEALTH RISKS.

THE TOBACCO INDUSTRY ACHIEVED THIS AIM BY FUNDING RESEARCH THAT SUGGESTED EXPLANATIONS FOR CANCER OTHER THAN SMOKING.
JOE SMOKER RIP
VAST SUMS OF MONEY WERE SPENT ON THIS CAUSE, A FIGURE THAT EXCEEDED $100 MILLION BY THE 1970s.
ANY RESEARCH THAT FAVORED ITS VIEWPOINT WAS PROMOTED, WHILE RESEARCH THAT DIDN'T WAS SUPPRESSED.
A WHOLE SERIES OF DISSENTING EXPERTS WERE PARADED IN ORDER TO BOLSTER THE ARGUMENT.
IN THIS WAY, TOBACCO FIRMS SUCCESSFULLY MUDDIED THE WATERS OF SCIENTIFIC RESEARCH,
WHILE GIVING THEMSELVES LEGAL MEANS BY WHICH TO OPPOSE REGULATION AND FIGHT COMPENSATION CLAIMS.

THESE TECHNIQUES HAVE BEEN ENTHUSIASTICALLY EMBRACED BY THE OIL AND GAS INDUSTRIES IN ORDER TO THROW DOUBT ON THE REALITY OF CLIMATE CHANGE.

THIS METHOD FAILS TO MAKE A DISTINCTION BETWEEN EVIDENCE AND OPINION
LET'S HAVE A HEATED DEBATE.
AND GIVES FREE PUBLICITY TO MARGINAL BELIEFS
GRR!
GRR!
SUCH AS ANTI-VACCINATIONISTS, HOMEOPATHY PROPONENTS,
BLAH BLAH BLAH!
AND CLIMATE CHANGE DENIERS.
MORE THAN NINETY PERCENT OF CLIMATE SCIENTISTS CONSIDER MAN-MADE CLIMATE CHANGE TO BE A REALITY.
HOWEVER, YOU WOULDN'T KNOW THIS FROM THE WAY THE ISSUE HAS BEEN PORTRAYED IN THE MEDIA.

THE MEDIA'S ATTEMPTS AT BALANCE HAVE FAR TOO OFTEN DISTORTED SCIENTIFIC DEBATE
AND THIS HAS HAD THE EFFECT OF MISLEADING THE GENERAL PUBLIC
IT'S SNOWING!
INTO THINKING THAT THERE ARE BASIC DIVISIONS AMONG SCIENTISTS WHEN THERE REALLY AREN'T.
SO MUCH FOR CLIMATE CHANGE.
JUST BECAUSE IT'S SNOWING NOW DOESN'T MEAN THE EARTH ISN'T WARMING UP.
ADVERSARIAL DISPUTE MIGHT BE AN EFFECTIVE WAY TO COVER POLITICS
OH YEAH!
BUT IT DOESN'T HELP AT ALL WHEN IT COMES TO EXPLAINING SCIENTIFIC ISSUES.

HUMANS ARE CHANGING THE CLIMATE. DENIERS DISAGREE, OF COURSE.

BUT THAT'S BECAUSE THEY INSIST ON HOLDING FIXED POSITIONS THAT HAVE NOTHING TO DO WITH SCIENCE.

THE SCIENTIFIC METHOD IS SELF-CORRECTING.
OBSERVATION.

THIS SELF-CORRECTION MAY TAKE TIME
QUESTIONS.
?

AND ANY RESULTS MAY BE MIRED IN CONTROVERSY UNTIL THE ISSUE IS SETTLED.
RESEARCH.

ONLY SCIENCE CAN REVEAL THE TRUE NATURE OF THE WORLD.
EXPERIMENT.

A SCIENTIST CAN'T AFFORD TO CHERRY-PICK DATA OR IGNORE INCONVENIENT FACTS
ANALYZE DATA.
BECAUSE ANY RESULTS WILL BE SUBJECT TO THE SAVAGE SCRUTINY OF OTHER SCIENTISTS
CONCLUSION.
I'VE GOT IT!
WHO WILL BE ONLY TOO KEEN TO PICK HOLES IN ANY THEORY.
THAT, THEN, IS MY CONCLUSION.
YOUR THEORY THAT THE MOON IS HELD IN PLACE BY AN INVISIBLE STRING IS QUITE ABSURD.
IF A THEORY IS TO SURVIVE, THEN ITS FINDINGS MUST BE REPEATABLE BY OTHERS
AND IT HAS TO BE ABLE TO BOTH EXPLAIN AND PREDICT EVENTS IN NATURE.

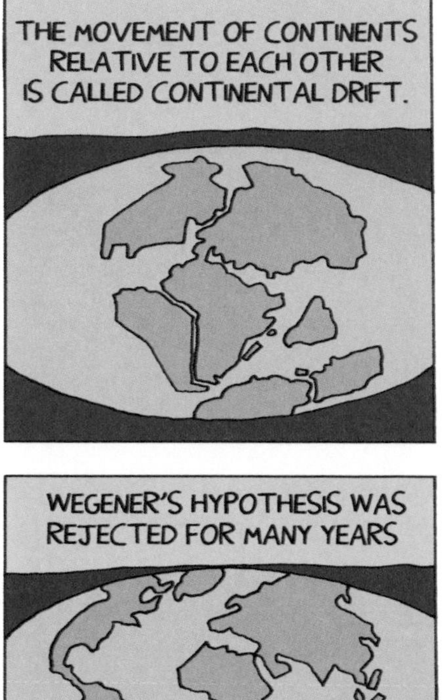
THE MOVEMENT OF CONTINENTS RELATIVE TO EACH OTHER IS CALLED CONTINENTAL DRIFT.
THIS THEORY WAS FIRST FULLY OUTLINED BY ALFRED WEGENER IN 1912.

WEGENER'S HYPOTHESIS WAS REJECTED FOR MANY YEARS

UNTIL THE THEORY OF TECTONIC PLATES, IN THE 1960s, GAVE AN EXPLANATION OF HOW SUCH MOVEMENT WAS POSSIBLE.

THIS THEORY FITS ALL OBSERVATIONS MADE AND IS NOW GENERALLY ACCEPTED.

OTHER THEORIES HAVE NOT FARED SO WELL.

ASTRONOMER FRED HOYLE'S STEADY STATE UNIVERSE THEORY IS A GOOD EXAMPLE OF THIS. HOYLE BELIEVED THAT MATTER WAS CONSTANTLY BEING CREATED BETWEEN THE GALAXIES, MAKING THE UNIVERSE EXPAND FOREVER.

AN INFINITE UNIVERSE, ENDLESS AND ETERNAL, THAT REQUIRED NO CAUSE AND NO CREATOR. AN ELEGANT THEORY, DESTROYED BY ANNOYING FACTS.

HOYLE'S THEORY WAS FATALLY DAMAGED BY THE DISCOVERY OF COSMIC BACKGROUND RADIATION. THE LAST ECHOES OF THE BIG BANG, PROOF OF A COSMIC STARTING POINT.

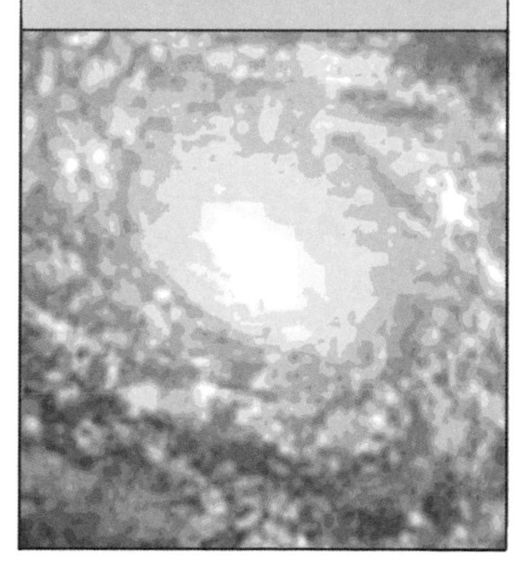

SCIENCE HAS A VERY ROBUST SYSTEM FOR ASSESSING THE QUALITY OF RESEARCH BEFORE IT'S PUBLISHED.
EXCUSE ME, BUT I NEED TO TAKE YOUR PAPER FOR REVIEW.
THIS SYSTEM IS CALLED PEER REVIEW.
NO! GO AWAY!
PEER REVIEW MEANS THAT OTHER EXPERTS IN THE SAME FIELD WILL CHECK RESEARCH PAPERS
WE HAVE TO CHECK IT.
FOR VALIDITY, SIGNIFICANCE, AND ORIGINALITY.
IT'S LIKE RUNNING A GAUNTLET.
EDITORS OF JOURNALS DRAW ON A LARGE POOL OF EXPERTS TO SCRUTINIZE PAPERS
DONT WORRY. WE'LL BE GENTLE.
BEFORE DECIDING WHETHER TO PUBLISH THEM.
OR RATHER WE WON'T. HAR!

MANY OF THE RESEARCH CLAIMS THAT APPEAR IN NEWSPAPERS, MAGAZINES, AND OTHER MEDIA
AND NOW THIS.
ARE NOT PUBLISHED IN PEER-REVIEWED JOURNALS.
POWER LINES ARE DANGEROUS.
IT'S WHY MANY REPORTED FINDINGS, SUCH AS CLAIMS ABOUT WONDER CURES OR NEW HEALTH SCARES
NEVER COME TO ANYTHING.
I NEVER HURT ANYONE.
AS A SOCIETY WE CAN'T BASE PUBLIC POLICY ON WORK THAT HASN'T BEEN ASSESSED AND COULD BE FLAWED.
PUBLICATION IN A PEER-REVIEWED JOURNAL IS ONLY THE FIRST STEP.

ANY FINDINGS AND THEORIES MUST GO ON TO BE RE-TESTED AGAINST OTHER WORK IN THE SAME FIELD.
MY PAPER'S BEEN ACCEPTED BY THIS JOURNAL.
SOME PAPERS' CONCLUSIONS WILL BE DISPUTED...
NOW THE SCRUTINY REALLY BEGINS.
WHILE FURTHER RESEARCH MAY SHOW THAT A PAPER NEEDS REVISION AS MORE DATA IS GATHERED.
BACK TO THE DRAWING BOARD.
THE PEER-REVIEW PROCESS CONNECTS LIKE-MINDED PEOPLE, LETTING THEM KNOW ABOUT NEW RESEARCH IN THEIR FIELD.
A SCIENTIST HAS TO HAVE THICK SKIN. SIGH!
IT'S A PERMANENT RECORD OF WHAT HAS BEEN DISCOVERED AND BY WHOM.
nature
HE LANCET
Science
IT HELPS SCIENTISTS PROMOTE THEIR WORK AND GAIN RECOGNITION FROM FUNDERS AND OTHER INSTITUTIONS.
TAKE THIS NOBEL PRIZE.

THERE ARE FLAWS IN THE PEER-REVIEW PROCESS.
IT CAN'T ALWAYS DETECT FRAUD OR MISCONDUCT. IF SOMEONE SETS OUT TO FALSIFY RESULTS. THEN THERE MAY BE NO WAY OF KNOWING THIS
UNTIL OTHERS TRY TO REPRODUCE THOSE RESULTS.
WAKEFIELD SACKED.
ANOTHER CRITICISM MADE IS THAT PEER REVIEW CAN BLOCK NEW IDEAS.
BUT, AS WE HAVE SEEN, IDEAS THAT CAN STAND CLOSE SCRUTINY WILL EVENTUALLY BE ACCEPTED.
SCIENCE WORKS.

AND IT WORKS IN A WAY THAT PSEUDO-SCIENCE SIMPLY CAN'T. IT HAS PASSED THE TESTS THAT OTHER FORMS OF THINKING HAVE FAILED.
CLICK
SCIENCE HAS BOTH BUILT THE TECHNOLOGICAL WORLD WE LIVE IN.
AND REVEALED TO US MANY OF THE DEEP MYSTERIES OF THE UNIVERSE.
HUH?
SCIENCE BUILDS AND ORGANIZES KNOWLEDGE IN THE FORM OF TESTABLE EXPLANATIONS AND PREDICTIONS.
SCIENCE IS THE MOST SUCCESSFUL TOOL EVER DEVISED FOR EXPLAINING OUR UNIVERSE.
THIS IS WHY SCIENCE PROCEEDS THE WAY IT DOES AND WHY IT IS SO POWERFUL.
END

SOURCES

In addition to the sources below, I also used a system not unlike the peer-review process we see with scientific papers. Each chapter was originally posted on my blog. This often led to a heated debate in the comments section; through this process I was able to clarify points I'd not properly understood, or make changes where I'd simply got things wrong. Many thanks to all the people who contributed in this process. —D.C.

The Moon Hoax

"Fox TV and the Apollo Moon Hoax." Bad Astronomy. Feb. 13, 2001 www.badastronomy.com/bad/tv/foxapollo.html (accessed Oct. 24, 2011).

Phillips, A. "The Great Moon Hoax." Nasa Science News. Feb. 23, 2001 http://science.nasa.gov/science-news/science-at-nasa/2001/ast23feb_2 (accessed Oct. 24, 2011).

Plait, P. *Bad Astronomy: Misconceptions and Misuses Revealed, from Astrology to the Moon Landing "Hoax."* John Wiley & Sons, 2002.

Shermer, M. "Fox Goes to the Moon, but NASA Never Did: The No-Moonies Cult Strike." 2001. http://homepages.wmich.edu/~korista/moonhoax2.html (accessed Nov. 13, 2011).

Homeopathy

Goldacre, B. "Newsnight/Sense About Science Malaria & Homeopathy Sting" *Bad Science*. Sept. 1, 2006. www.badscience.net/2006/09/newsnightsense-about-science-malaria-homeopathy-sting-the-transcripts.

"Homeopathy: What's the Harm?" *10:23*. www.1023.org.uk/whats-the-harm-in-homeopathy.php (accessed Oct. 24, 2011).

Hope, A., (State Coroner). "Coronial Inquest into the Death of Penelope Dingle." www.safetyandquality.health.wa.gov.au/docs/mortality_review/inquest_finding/Dingle_Finding.pdf (accessed Oct. 24, 2011).

Richmond, C. "Jacques Benveniste; Maverick Scientist Behind a Controversial Experiment into the Efficacy of Homeopathy." *Guardian*, Oct. 21, 2004. www.guardian.co.uk/science/2004/oct/21/obituaries (accessed Oct. 24, 2011).

Singh, S. "Could Water Really Have a Memory?" *BBC News*. Jul. 25, 2008 http://news.bbc.co.uk/1/hi/health/7505286.stm (accessed Oct. 24, 2011).

"What Is Homeopathy?" *10:23*. www.1023.org.uk/what-is-homeopathy.php (accessed Oct. 24, 2011).

Chiropractic

Adams, C. "Is Chiropractic for Real or Just Quackery?" *The Straight Dope*. Jun. 2, 2008. www.straightdope.com/columns/read/2771/is-chiropractic-for-real-or-just-quackery (accessed Nov. 13, 2011).

Hansen, J. "Doctors Accuse Chiropractors of Selling Anti-vaccination Message." *News.com.au*. July 27, 2011. www.news.com.au/national/doctors-accuse-chiropractors-of-selling-anti-vaccination-message/story-e6frfkvr-1226102836863 (accessed Nov. 13, 2011).

Novella, S. "Chiropractic—a Brief Overview." *Science-Based Medicine.* Jun. 24, 2009 www.sciencebasedmedicine.org/index.php/chiropractic-a-brief-overview-part-i (accessed Nov. 13, 2011).
Singh, S. "Beware the Spinal Trap." *Guardian.* Apr. 19, 2008. www.guardian.co.uk/commentisfree/2008/apr/19/controversiesinscience-health (accessed Nov. 13, 2011).
Singh, S. and Ernst E. *Trick or Treatment? Alternative Medicine on Trial.* Bantam Press, 2008.
"What Alternative Health Practitioners Might Not Tell You." *ebm-first.* Oct. 5, 2011. www.ebm-first.com/chiropractic/risks.html (accessed Nov. 13, 2011).
Wolff, J. "Deadly Twist: Neck Adjustments Can Be Risky." *MSNBC.* Jun. 17, 2007. www.msnbc.msn.com/id/18871755/#.Trlwrxzx-fl (accessed Nov. 13, 2011).

The MMR Vaccination Scandal

Boseley, S. "Andrew Wakefield Found 'Irresponsible' by GMC Over MMR Vaccine Scare." *Guardian* Jan. 28., 2010 www.guardian.co.uk/society/2010/jan/28/andrew-wakefield-mmr-vaccine (accessed Oct. 24, 2011).
Deer, B. "How the Case Against the MMR Vaccine Was Fixed." *BMJ* 2011; 342:c5347. Jan. 5, 2011. www.bmj.com/content/342/bmj.c5347.full (accessed Oct. 24, 2011).
Godlee, F., Smith, J., and Marcovitch, H. "Wakefield's Article Linking MMR Vaccine and Autism Was Fraudulent. " *BMJ* 2011; 342:c7452. 2011 Jan 5. www.bmj.com/content/342/bmj.c7452.full (accessed Oct. 24, 2011).
Goldacre, G. *Bad Science.* Fourth Estate, 2008.
Gorski, D. "The Fall of Andrew Wakefield." *Science-Based Medicine.* Feb. 22, 2010. www.sciencebasedmedicine. org/index.php/the-fall-of-andrew-wakefield (accessed Oct. 24, 2011).
"The Lancet Scandal." http://briandeer.com/mmr-lancet.htm (accessed Oct. 24, 2011).
"MMR Doctor Given Legal Aid Thousands." *Sunday Times.* Dec. 31, 2006. http://briandeer.com/mmr/st-dec-2006.htm (accessed Oct. 24, 2011).
"The Media's MMR Hoax." *Bad Science.* Aug. 30, 2008. www.badscience.net/2008/08/the-medias-mmr-hoax (accessed Oct. 24, 2011).
"Solved—the Riddle of MMR." http://briandeer.com/solved/solved.htm (accessed Oct. 24, 2011).
"The Wakefield Factor." http://briandeer.com/wakefield-deer.htm (accessed Oct. 24, 2011).

Evolution

Coyne, J. A. *Why Evolution Is True.* Oxford University Press, 2009.
Dawkins, R. *The Selfish Gene: 30th Anniversary Edition.* Oxford University Press, 2006.
Gould, S. J. *Hen's Teeth and Horse's Toes: Further Reflections on Natural History.* Penguin, 1990.
Grant, S. "Fine Tuning the Peppered Moth Paradigm." *Evolution* 1999. 53(3), p. 980–84.
"Is Evolution a Theory or a Fact?" National Academy of Sciences and Institute of

Medicine. http://nationalacademies.org/evolution/TheoryOrFact.html (accessed Nov. 8, 2011).
Jones, S. *Almost Like a Whale: The Origin of Species Updated.* Ballantine, 1999.
Shubin, N. *Your Inner Fish: The Amazing Discovery of Our 375-Million-Year-Old Ancestor.* Penguin, 2009.
Weiner, J. *The Beak of the Finch: A Story of Evolution in Our Time*. Vintage, 1995.

Fracking

"Debunking Gasland." *Energy in Depth*. Jun. 9, 2010. http://www.energyindepth.org/debunking-gasland (accessed May 27, 2012).
Dunning, B. "All About Fracking." Skeptoid Podcast. Skeptoid Media, Inc., Sept. 13, 2011; Aug. 30, 2012 (web). http://skeptoid.com/episodes/4275 (accessed May 26, 2012).
Gasland: A Film by Josh Fox (2009). International WOW Company. http://www.gaslandthemovie.com (accessed May 27, 2012).
"The Halliburton Loophole" (editorial), *New York Times*, Nov. 2, 2009. http://www.nytimes.com/2009/11/03/opinion/03tue3.html?_r=1 (accessed May 26, 2012).
"Injection Wells: The Poison Beneath Us." ProPublica. June 12, 2012. http://www.propublica.org/article/injection-wells-the-poison-beneath-us (accessed May 15, 2012).
Kimball, J. "Congress Releases Report on Toxic Chemicals Used in Fracking." 8020 Vision. Apr. 7, 2011. http://8020vision.com/2011/04/17/congress-releases-report-on-toxic-chemicals-used-in-fracking/ (accessed May 15, 2012).
Lustgarten, A. "EPA Finds Compound Used in Fracking in Wyoming Aquifer." ProPublica, Nov. 10, 2011. http://www.propublica.org/article/epa-finds-fracking-compound-in-wyoming-aquifer (accessed May 5, 2012).
"Regulation Lax as Gas Wells' Tainted Water Hits Rivers." *New York Times*. Feb. 26, 2011. http://www.nytimes.com/2011/02/27/us/27gas.html?pagewanted=all (accessed May 27, 2012).
U.S. Environmental Protection Agency (EPA). "Hydraulic Fracturing Background Information." http://water.epa.gov/type/groundwater/uic/class2/hydraulicfracturing/wells_hydrowhat.cfm (accessed May 26, 2012).
Urbina, I. "E.P.A. Steps Up Scrutiny of Pollution in Pennsylvania Rivers." *New York Times*. Mar. 7, 2011. http://www.nytimes.com/2011/03/08/science/earth/08water.html?_r=1 (accessed May 15, 2012).
Williams, D. "Frequent Pit Liner Leaks Argue Against Hickenlooper Call for Less Regulation." *Colorado Independent*. Jun. 28, 2010. http://coloradoindependent.com/56391/frequent-pit-liner-leaks-argue-against-hickenlooper-call-for-less-regs (accessed May 27, 2012).

Climate Change

Carey, B. "January to Be Coldest Since 1985, Where's Global Warming?" *Live Science*. Jan. 7, 2011. www.livescience.com/9227-january-coldest-1985-global-warming.html (accessed Nov. 13, 2011).
Carrington, D. "Q&A: 'Climategate'." *Guardian*. Jul. 7, 2010. www.guardian.co.uk/environment/2010/jul/07/climate-emails-question-answer (accessed Nov. 13, 2011).

Cook, J. "10 Indicators of a Human Fingerprint on Climate Change." *Skeptical Science*. Jul. 30, 2010. www.skepticalscience.com/10-Indicators-of-a-Human-Fingerprint-on-Climate-Change.html (accessed Nov. 13, 2011).

Dow, K. and Downing, T. E. *The Atlas of Climate Change*. Earthscan, 2006.

Mandia, S. "Global Warming: Man or Myth? Climategate Coverage: Unfair & Unbalanced." http://profmandia.wordpress.com/ 2010/04/18/climategate-coverage-unfair-unbalanced (accessed Oct. 24, 2011).

Mayer, J. "Covert Operations: The Billionaire Brothers Who Are Waging a War Against Obama." *New Yorker*. Aug. 30, 2010. www.newyorker.com/reporting/2010/08/30/100830fa_fact_mayer? currentPage=all (accessed Nov. 13, 2011).

Monbiot, G. "The Climate Denial Industry Is Out to Dupe the Public. And It's Working." *Guardian*. Dec. 7, 2009. www.guardian.co.uk/commentisfree/cif-green/2009/dec/07/climate-change-denial-industry (accessed Nov. 13, 2011)

Oliver, C., Frigieri, G., and Clark, D. "Everything You Need to Know About Climate Change—Interactive." *Guardian*. Aug. 15, 2011. www.guardian.co.uk/environment/interactive/2011/aug/15/everything-know-climate-change (accessed Nov. 13, 2011).

Oreskes, N. and Conway, E. M. *Merchants of Doubt: How a Handful of Scientists Obscured the Truth on Issues from Tobacco Smoke to Global Warming*. Bloomsbury, 2010.

Science Denial

Beresford, D. "Manto Tshabalala-Msimang" (obituary). *Guardian*. Jan. 7, 2010. www.guardian.co.uk/world/2010/jan/07/manto-tshabalala-msimang-obituary (accessed Nov. 13, 2011).

Boseley, S. "Mbeki Aids Denial 'caused 300,000 deaths'." *Guardian*. Nov. 26, 2008. www.guardian.co.uk/world/2008/nov/26/aids-south-africa (accessed Nov. 13, 2011).

Eriksen, M., Mackay, J. and Shafey, O. *The Tobacco Atlas*. 2nd ed. American Cancer Society, 2006.

Foucart, S. "When Science Is Hidden Behind a Smokescreen." *Guardian*. Jun. 28, 2011. www.guardian.co.uk/science/2011/jun/28/study-science-research-ignorance-foucart? (accessed Nov. 13, 2011).

Goodell, J. "As the World Burns: How Big Oil and Big Coal Mounted One of the Most Aggressive Lobbying Campaigns in History to Block Progress on Global Warming." *Rolling Stone*. Jan. 6, 2010. www.rollingstone.com/politics/news/as-the-world-burns-20100106 (accessed Nov. 13, 2011).

Harris, W. "How the Scientific Method Works." *How Stuff Works*. http://science.howstuffworks.com/innovation/scientific-experiments/scientific-method2.htm (accessed Nov. 13, 2011).

McKee, M. "Denialism: What Is It and How Should Scientists Respond?" *European Journal of Public Health* 2009 vol. 9, nos. 1, 2–4.

McKie, R. "Fred Hoyle: the Scientist Whose Rudeness Cost Him a Nobel Prize." *Guardian*. 2010 Oct. 3, 2010. www.guardian.co.uk/science/2010/oct/03/fred-hoyle-nobel-prize?INTCMP=ILCNETTXT3487 (accessed Nov. 13, 2011).

"Science and Truth Have Been Cast Aside by Our Desire for Controversy." *Guardian*. Jul. 24, 2011. www.guardian.co.uk/commentisfree/2011/jul/24/science-reporting-climate-change-sceptics (accessed Nov. 13, 2011).